Lecture Notes on Data Engineering and Communications Technologies

Volume 154

Series Editor

Fatos Xhafa, Technical University of Catalonia, Barcelona, Spain

The aim of the book series is to present cutting edge engineering approaches to data technologies and communications. It will publish latest advances on the engineering task of building and deploying distributed, scalable and reliable data infrastructures and communication systems.

The series will have a prominent applied focus on data technologies and communications with aim to promote the bridging from fundamental research on data science and networking to data engineering and communications that lead to industry products, business knowledge and standardisation.

Indexed by SCOPUS, INSPEC, EI Compendex.

All books published in the series are submitted for consideration in Web of Science.

Eric C. K. Cheng · Tianchong Wang ·
Tim Schlippe · Grigorios N. Beligiannis
Editors

Artificial Intelligence in Education Technologies: New Development and Innovative Practices

Proceedings of 2022 3rd International Conference on Artificial Intelligence in Education Technology

Editors
Eric C. K. Cheng
Department of Curriculum and Instruction
The Education University of Hong Kong
Tai Po, Hong Kong

Tim Schlippe
IU International University of Applied Sciences
Erfurt, Germany

Tianchong Wang
Swinburne University of Technology
Melbourne, VIC, Australia

Grigorios N. Beligiannis
Department of Food Science & Technology
University of Patras
Agrinio, Greece

ISSN 2367-4512 ISSN 2367-4520 (electronic)
Lecture Notes on Data Engineering and Communications Technologies
ISBN 978-981-19-8039-8 ISBN 978-981-19-8040-4 (eBook)
https://doi.org/10.1007/978-981-19-8040-4

© The Editor(s) (if applicable) and The Author(s), under exclusive license to Springer Nature Singapore Pte Ltd. 2023
This work is subject to copyright. All rights are solely and exclusively licensed by the Publisher, whether the whole or part of the material is concerned, specifically the rights of translation, reprinting, reuse of illustrations, recitation, broadcasting, reproduction on microfilms or in any other physical way, and transmission or information storage and retrieval, electronic adaptation, computer software, or by similar or dissimilar methodology now known or hereafter developed.
The use of general descriptive names, registered names, trademarks, service marks, etc. in this publication does not imply, even in the absence of a specific statement, that such names are exempt from the relevant protective laws and regulations and therefore free for general use.
The publisher, the authors, and the editors are safe to assume that the advice and information in this book are believed to be true and accurate at the date of publication. Neither the publisher nor the authors or the editors give a warranty, expressed or implied, with respect to the material contained herein or for any errors or omissions that may have been made. The publisher remains neutral with regard to jurisdictional claims in published maps and institutional affiliations.

This Springer imprint is published by the registered company Springer Nature Singapore Pte Ltd.
The registered company address is: 152 Beach Road, #21-01/04 Gateway East, Singapore 189721, Singapore

Conference Committee

Conference Chair

Xinguo Yu   Central China Normal University, China and University of Wollongong, Australia

Conference Co-chair

Ben du Boulay   University of Sussex, UK

Program Chairs

Eric C. K. Cheng   The Education University of Hong Kong, Hong Kong, China
Yanjiao Chen   Zhejiang University, China

Program Co-chairs

Tim Schlippe   IU International University of Applied Sciences, Germany
Grigorios N. Beligiannis   University of Patras, Greece

Publication Chair

Tianchong Wang   Swinburne University of Technology, Australia

Technical Program Committee

P. K. Paul	Raiganj University, India
Teh Faradilla Abdul Rahman	Univevrsiti Teknologi MARA, Malaysia
Yu-Mei Wang	University of Alabama at Birmingham, USA
Loc Nguyen	International Engineering and Technology Institute (IETI), China
Xinghong Liu	Hubei Normal University, China
Norma Binti Alias	Technology University of Malaysia, Malaysia
Francesco Flammini	Mälardalen University, Sweden
Jia Chen	Central China Normal University, China
Mao Feng	Shanghai University of International Business and Economics, China
Wang Aiwen	Guangzhou City Construction College, China
Bin Xue	National University of Defense Technology, China
Mauro Gaggero	National Research Council of Italy, Institute of Marine Engineering, Italy
Mohd Faisal Hushim	Universiti Tun Hussein Onn, Malaysia
Dariusz Jacek Jakóbczak	Technical University of Koszalin, Poland
Xinhua Xu	Hubei Normal University, China
Lei Niu	Central China Normal University, China

Preface

Rapid developments in artificial intelligence in education technology and its appalling potential have drawn significant attention in recent years. AIET 2022 establishes a platform for researchers in the field of artificial intelligence in education technology to present research, exchange innovative ideas, propose new models, and demonstrate advanced methodologies and novel systems.

This proceeding is a compendium of selected research papers presented at the 2022 3rd International Conference on Artificial Intelligence in Education Technology (AIET 2022)—held virtually during July 1–3, 2022. Providing broad coverage of recent technology-driven advances, the proceeding is an informative and valuable resource for researchers, practitioners, education leaders, and policy-makers who are involved or interested in artificial intelligence in education technology.

Prestigious experts and professors have been invited as keynote speakers and invited speakers to deliver the latest information in their respective expertise areas. It will be a golden opportunity for students, researchers, and engineers to interact with the experts in their fields of research.

On behalf of the Organizing Committee, we would like to thank the authors for the high quality of the submissions. We are grateful to all reviewers for being responsive and thorough in such a short time frame. The creation of such a broad and high-quality conference program would not have been possible without their involvement. We are grateful to the members of the AIET 2022 conference committees for their work to make this professional conference a success.

Xinguo Yu
Conference Chair
Central China Normal University
Wuhan, China

University of Wollongong
Wollongong, Australia

Contents

Machine Learning and Data Analysis in Education

Towards Effective Teacher Professional Development for STEM
Education in Hong Kong K-12: A Case Study 3
Tianchong Wang and Eric C. K. Cheng

Identification and Hierarchical Analysis of Risk Factors in Primary
and Secondary Schools: A Novel GT-DEMATEL-ISM Approach 20
Shuzhen Luo, Qian Wang, and Peirong Qi

Generation of Course Prerequisites and Learning Outcomes Using
Machine Learning Methods 34
Polina Shnaider, Anastasiia Chernysheva, Maksim Khlopotov,
and Carina Babayants

Learning Factors for TIMSS Math Performance Evidenced
Through Machine Learning in the UAE 47
Ali Nadaf, Samantha Monroe, Sarath Chandran, and Xin Miao

Educational Information Technology and E-Learning

Explainability in Automatic Short Answer Grading 69
Tim Schlippe, Quintus Stierstorfer, Maurice ten Koppel,
and Paul Libbrecht

The Framework Design of Intelligent Assessment Tasks
Recommendation System for Personalized Learning 88
Qihang Cai and Lei Niu

Assessing Graduate Academic Scholarship Applications
with a Rule-Based Cloud System 102
Yongbin Zhang, Ronghua Liang, Yuansheng Qi, Xiuli Fu,
and Yanying Zheng

**AI-Based Visualization of Voice Characteristics in Lecture Videos'
Captions** 111
Tim Schlippe, Katrin Fritsche, Ying Sun, and Matthias Wölfel

**The Intergroup Bias in the Effects of Facial Feedback
on the Recognition of Micro-expressions** 125
Kunling Peng, Yaohan Wang, and Qi Wu

**Feedback on the Result of Online Learning of University Students
of Health Sciences** 135
Carmen Chauca, Ynés Phun-Pat, Maritza Arones,
and Olga Curro-Urbano

Educational Management, Psychology and Educational Statistics

**Exploration and Application of the Blended Learning Model
in the "Software Engineering" Course** 147
Mengmei Wang

**Improving Learning Outcomes with Pair Teaching StrateFiggy
in Higher Education: A Case Study in C Programming Language** 157
Yongbin Zhang, Ronghua Liang, Yuansheng Qi, Xiuli Fu,
and Yanying Zheng

**Foreign Language Reading Anxiety and Its Correlation
with Reading Test Scores** 168
Van T. T. Dang and Trung Nguyen

**The Relationship between Nostalgia and Life Satisfaction
in College: A Chained Mediation Model** 182
Daosheng Xu and Yiwen Chen

**Reform and Practice of Talent Training Model Based on Cold
Chain Industry College** 193
Huichuan Dai, Huihua Shang, and Yefu Tang

Study on the Growth Pattern of Middle-Level Vocational Skills 203
Xinqiang Meng, Le Qi, and Mengyang Liu

Author Index 217

Machine Learning and Data Analysis in Education

Towards Effective Teacher Professional Development for STEM Education in Hong Kong K-12: A Case Study

Tianchong Wang[1] and Eric C. K. Cheng[2]

[1] Department of Education, Swinburne University of Technology, Hawthorn 3122, Australia
tianchongwang@swin.edu.au
[2] Department of Curriculum and Instruction, The Education University of Hong Kong, Taipo, N.T., Hong Kong S.A.R., China

Abstract. STEM education has taken on high importance in Hong Kong K-12 education landscape. Despite policy advocacy and curriculum endeavour, the quality of STEM education varies significantly between schools. Research literature indicates that high-quality STEM education requires teachers' rigorous delivery of topics and appropriate pedagogies, and one approach to improve such practices is teacher professional development (TPD). However, because current research on TPD has not given explicit consideration to the complex nature of STEM education, there remains a lack of a clear blueprint of how TPD should be conducted to build teachers' capacity for STEM education effectively. This paper presents a case study that explores the necessary attributes and identifies the missing links of STEM education TPD by understanding how various TPD models supported a Hong Kong K-12 school embracing STEM education. Qualitative data collection methods, including semi-structured interviews and classroom observations, were employed to draw a picture of TPD implementations in the selected school. The findings suggest that, at the macro-level, effective STEM TPD should not stop at employing mixed use of TPD models; the models have to be integrated organically with respect to a school based STEM curriculum implementation approach. A collaborative culture between teachers must be cultivated for effective interdisciplinary integration. Collaborative action research should also be promoted to develop collective wisdom of STEM pedagogies. At the micro-level, TPACK and cross-disciplinary integration skills need to be focusing areas of STEM TPD. With these guiding principles, some possible strategies for effective STEM education TPD are suggested.

Keywords: STEM education · Teacher professional development · TPD models · K-12

1 Background

The increasing sophistication of new technologies such as artificial intelligence (AI) has generated anxieties about the future of work and encouraged reflections about how we can equip our students to thrive in an interconnected and technology-rich world (OECD 2018). As a result, a widespread consensus has been reached over the strategic importance of promoting STEM education (Freeman et al. 2014), which is believed to be necessary for preparing the future workforce to meet demands for the emerging *Industrial 4.0* (Schwab 2017). In many education systems around the globe, STEM education has become a priority, and the "*Smart City*" (Central Policy Unit 2015) Hong Kong is no exception to being one of the economies that aggressively promote STEM education.

The introduction of STEM education in Hong Kong started with policy advocacy and curriculum endeavour. The notion was presented in the *2015 Government Policy Address* (Government of the Hong Kong Special Administrative Region 2015) as an essential education reform measure. In addition to adding STEM as one of the Key Learning Areas (KLAs) in the *Learning to Learn 2.0+* (Curriculum Development Council 2015), the city's Education Bureau (EDB) has been serving as the key driving force in promoting this mandate. More specifically, EDB's *The Report on Promotion of STEM Education: Unleashing Potential in Innovation* (Education Bureau of Government of HKSAR 2016) calls for holistic strategies to nurture school students' creativity, collaboration, innovation, and problem-solving skills in STEM; particularly, student-centred pedagogies are highly encouraged to facilitate integrative problem-solving skills, learning to code, and entrepreneurial spirit. The report also encourages schools to take their initiatives and pursue a cross-subject approach to implementing STEM education in the school-based curriculum.

Despite the heightened attention and commitment from the government level, infusing STEM education in Hong Kong K-12 education still faces several challenges at the school level. First, there is a general lack of consensus between schools on how STEM education should be positioned in the school-based curriculum—probably due to the disputed interpretations of its nature and purposes (Kelley and Knowles 2016; Margot and Kettler 2019). Second, whether schoolteachers have the technical, content, and pedagogical knowledge (TPACK) (Koehler et al. 2013) needed to teach is also a challenge in implementing STEM education in K-12 classes (Chai 2019). Current teachers who have received teacher training in only one subject area may be unable to adopt an integrated and holistic approach to teaching STEM (Aslam et al. 2018; Margot and Kettler 2019). Third, as it is debatable between schools whether STEM education should be implemented as a separate subject, cross-curricular or even extra-curricular practice, the pedagogy of STEM education could vary significantly. As a result, teachers have yet to develop a clear vision for teaching STEM education. These challenges have subsequently led to quality issues in STEM education in Hong Kong K-12 schools, hindering the outcomes of this well-intended education reform agenda.

Literature suggests that teacher professional development (TPD) is a powerful instrument that schools use to support teachers to address challenges and continue to strengthen their practice (Darling-Hammond et al. 2017). However, there are only a few research studies on TPD for STEM (Chai 2019), and recommendations made by these studies (e.g., Brenneman et al. 2019) may have limited referencing value for Hong Kong K-12 as they are often context-bound. Moreover, current studies on TPD models and facilitating strategies are out of step with STEM education. Many have not given explicit consideration to the complex nature of STEM education, yet this factor is critical for ensuring the effectiveness of its TPD.

Our study, therefore, seeks to enrich the literature by identifying necessary attributes and missing links and subsequently proposes guiding principles for effective STEM education TPD. We believe exploring the necessary attributes and missing links for effective TPD for STEM education would have immediate implications for successful STEM education integration into Hong Kong K-12.

The rest of this paper is structured into four parts. In Sect. 2, the study is framed with an overview of STEM Education and a review of TPD models. The research questions, the case study research design, and data collection methods are detailed in Sect. 3. The fourth section of the paper presents findings and discussions from our case study investigation. Finally, the implications of what we have learned and the concluding remarks are offered.

2 Literature Review

2.1 From *Education for STEM* to *STEM Education*

The term STEM was coined by the National Science Foundation (NSF) of the United States in the late 1990s as an overarching acronym covering Science, Technology, Engineering and Mathematics—the four discipline areas that provide the "fuel" necessary for powering the rapidly developing world today. The four disciplines have long been taught at the primary and secondary levels, with different levels of emphasis and extension. Whether integrated or not, the education for STEM aims to expand a STEM-literate population and create a talent pool of STEM professionals so that the country would not fall behind in global economic competitiveness (Sanders 2009).

A more evolved conceptualisation of *STEM education* has emerged in recent years (Pickering et al. 2016). In contrast to the previous practices, STEM education draws on the interrelationships among these disciplines and brings them together, as it is believed that learning through multiple, integrated subjects can produce deeper conceptual understandings, better development of skills, and better achievement than learning the subjects in isolation (Moore et al. 2015). In line with such thinking, STEM education has been, however, exercised in diversified forms. This is understandable because there are considerable variations in what constitutes STEM education (Breiner et al. 2012; Lamberg and Trzynadlowski 2015). For example, some scholars such as Bybee (2013), Sanders (2009), and Tytler et al. (2015) considered STEM education as a spectrum of learning activities built around authentic problems which involve some or all of the respective disciplines—S, T, E and M—and have an interdisciplinary nature at its core. Other scholars such as Merrill and Daugherty (2009) saw STEM education as a meta-discipline based on learning standards where teaching has integrated teaching and learning approaches, and where specific content is undivided, contemplating a dynamic and fluid instruction. Recently, AI education is also considered part of STEM education by many (Wang and Cheng 2021, 2022).

Although it may not be necessary or even feasible to coalesce around one standard definition of STEM education, some scholars have found that the notion of STEM education is not without some shared understandings (Han et al. 2015). Regardless of how it is interpreted, STEM education is not simply a new name for the traditional approach to teaching science and mathematics. Nor is it just the grafting of "technology" and "engineering" layers onto standard science and math curricula. Instead, STEM education is an approach to teaching larger than its constituent parts, notably in developing students' generic skills such as design thinking (Cross 2011) and computational thinking (Wing 2006). Design thinking comprises the stages of empathising, defining, ideating, prototyping, and testing (Cross 2011). Rather than merely being an integral part of the process in engineering and technology design, STEM education researchers consider design thinking as a general cognitive process that can take place in many different fields (Razzouk and Shute 2012), and such a mentality has to be nurtured for every student (Strimel et al. 2018, 2019). Similarly, computational thinking (Wing 2006), originally a concept in the Computer Science field, represents an imperative thinking skill set that everyone should master: abstraction and decomposition, thinking recursively, problem reduction and transformation, error prevention and protection, and heuristic reasoning which are needed to solve universal complex problems.

While advocating design thinking and computational thinking is clear for STEM education (Chai et al. 2020a), how the learning and teaching practices should be done is still being explored by the STEM education research community at large. Among these studies, one of the most common issues reported is the scarcity of teachers' STEM education knowledge, particularly the insufficiency of TPACK (e.g., Chai 2019; Chai et al. 2019, 2020b) and the lack of curriculum integration skills (e.g., Honey et al. 2014). As such, many STEM education studies urged using TPD as a critical strategy for improvement.

2.2 Teacher Professional Development

Teachers are responsible for implementing the curriculum in the classroom. They play a pivotal role in engaging students in their learning, supporting them in monitoring and managing their own learning, and enabling them to enhance their learning outcomes. While teachers have the requisite competencies to carry out such a role, they need to keep their knowledge and skills relevant and up-to-date; no matter how comprehensive the pre-service training teachers have received, it cannot be expected that teachers can cope with all the new changes and new challenges they will face throughout their careers (OECD 2009). This need has been particularly intensified for teachers involving STEM education; the majority of the current teachers would find it challenging to adopt an integrated and holistic approach to teaching STEM as they only received teacher education in one subject area (Aslam et al. 2018).

Schools, therefore, often encourage teachers' participation in specific programmes, commonly known as TPD, beyond their initial training (Darling-Hammond 2006). TPD is defined as "those processes and activities designed to enhance the professional knowledge, skills and attitudes of educators so that they might, in return, improve learning of students (Guskey 2000, p. 16). As a complex system, TPD is multi-causal, multi-dimensional and multi-quarrelational (Harwell 2003; Opfer and Pedder 2011). Besides the learning activities per se, TPD relates to teachers at both individual and school levels. At the individual level, successful TPD requires cognitive and emotional involvement of teachers, the capacity and willingness to examine where each one stands in terms of convictions and beliefs, and the perusal and enactment of appropriate alternatives for improvement or change (Avalos 2011). At the school level, principal support, collaborative learning culture, and the school's TPD policy were identified as enabling factors for success (Cheng 2017; Hords 1997). Among these factors, principal support was highlighted as a critical factor because school leadership could inspire in teachers a vision of TPD, and it could alert teachers to be aware of how TPD can contribute to the competitive performance and sustainable development of a school towards an envisioned future (Cheng 2015). Principal support was also identified as having a predictive effect on TPD policy and collaborative learning culture, while the effectiveness of a TPD plan is predicted by a collaborative culture and management strategy (Cheng 2017). Cheng and Ko (2012) found that principals could manage TPD activities strategically by formulating policies to link the objectives of TPD with the school development plan. Placing TPD in the school development plan is necessary for achieving its successful implementation. In addition, open culture in the schools is also identified as an important factor for a significant long-term effect on teachers' competencies and professional orientation (Gaikhorst et al. 2017). Organisational learning culture enables professional dialogues among teachers, which enhance the continual enhancement of collective capacities and the improvement of team effectiveness (Senge 2012). To encourage such a culture for TPD, trust between teachers and school leadership is a necessity (Cheng 2018).

TPD can be structured and organised in multiple ways. Each TPD model has its own strengths and weaknesses, as reviewed in the following. The typology of models provided us with a lens to examine a school's TPD practices and therefore served as the conceptual framework that guided this study.

The Training Model. The training model has arguably been the dominant form of TPD (Hooker 2008). Signified by a top-down approach, this model supports a skills-based, technocratic view of teaching whereby TPD provides teachers with the opportunity to update their skills as a means to demonstrate their competence. Regardless of being subject-based or interest-based, the training is generally delivered to the teacher by an expert, and the participant has a passive role. This model of TPD is often subject to criticism for its lack of sufficient connection to the current classroom context in which teachers work. In such a model, *"teachers are knowledge users, not generators"* (Cochran-Smith and Lytle 1999, p. 257).

The Cascade Model. The cascade model of TPD typically involves taking a few selected teachers out of their school environment to a centralised location. The selected teachers are trained together in a 'training of trainers', and then return to their own schools. The trained teachers then disseminate what they have learned to the remaining teachers in their schools. This process can be facilitated with the help of online technologies such as MOOCs (Laurillard et al. 2018; Laurillard and Kennedy 2019). The cascade model is often used as it offers a way of reaching many teachers within a short space of time, and is cost-effective (Hardman 2011; Hayes 2000; Ono and Ferreira 2010). However, this model also has several flaws, one of the biggest being the risk of dilution of quality as the knowledge or skills are passed down the line (Turner et al. 2017). Similar to the training model, the cascade model only supports a technicist view of teaching, where skills and knowledge are given priority over attitudes and values; what is passed on in the cascading process generally addresses questions of "what" and "how" but rarely questions of "why" (Kennedy 2005).

Action Research Model. Action research (Elliott 1976) is a form of self-reflective enquiry conducted by participants to improve their educational practices, their understanding of these practices and the situations in which the practices are carried out (Kemmis 1988). Action research "typically involves small-scale investigative projects in the teacher's own classroom, and consists of several phases which often recur in cycles: planning, action, observation, and reflection" (p. 12). The process of action research has proven to be a powerful tool for TPD (Zeichner 2003). In the action research process, teachers are expected to learn cooperatively and become reflective practitioners (Schon 1983) by applying theories postulated by others. Adopting action research into TPD could provoke critical reflection on their beliefs and conceptions about the role of teachers, teaching and learning, therefore educates reflective teachers to deal with the complexity of practice (Guskey 2002; Mills 2007). Brown et al. (2015) also claim that a model of TPD that is located in a strong orientation to reflexive practice and focus on understanding learners together may have facilitated positive changes in the teachers' practices.

Classroom/school-based Partnership Model. Ufnar and Shepherd (2019) propose a classroom partnership model to support K-12 teachers' TPD for STEM education. They conducted a study to determine the core, structural features of a Scientist in the Classroom Partnership program necessary for effective teacher PD. They found that the participant gains discipline and pedagogical content knowledge, inquiry strategies, and renewal of teaching. Cheng (2018) portrays a Japanese Lesson Study model as a kind of classroom partnership model to support TPD in pedagogical knowledge through a plan-do-check-act cycle for instructional design. Svendsen (2016) claimed that school-based collaborative TPD programmes positively impact their teaching practices, ways of thinking about teaching, and attitudes towards collaboration.

3 Methodology

3.1 Research Questions

This study aims to identify the necessary attributes and missing links toward effective TPD for STEM education by exploring the promising practices and lessons learned from a Hong Kong K-12 school's journey of STEM education introduction with the help of TPDs. We attempted to answer the following questions:

1. *How are the TPD models implemented as the selected Hong Kong K-12 school pursuing STEM education?*
2. *What are lessons learned as the TPD models are being employed?*

3.2 Research Design

This study applied an exploratory research design (Creswell 2014) based on the nature of our research questions. The qualitative case study (Yin 2003) methodology was implemented as it enabled the researchers to understand real-life phenomena through the actor's perspective in closely examining the data within a specific context. This demanded a detailed, in-depth data collection involving multiple sources of information (Creswell 2007; Marshall and Rossman 2011). In this study, we incorporated multiple qualitative data collection methods, and the researchers were the key instruments for data collection. Applying more than one method of data collection in the study was intended to provide a deeper understanding of the case and an exploration of the experience from a range of perspectives when seeking to answer the research questions (Merriam 2009). A total of eleven semi-structured interviews were conducted with the principal, curriculum leaders and teachers involved in STEM education at School A. In line with the research questions, the guiding questions of the interviews covered: (a) perspectives on STEM education, (b) approaches to TPD for STEM education, and (c) opportunities and challenges involved in TPD. Concurrently, a total of five STEM education lessons were observed to triangulate the data, and at the same time, draw a more comprehensive portrait of STEM education TPD implementations at School A.

3.3 Case Selection and Research Settings

This case study involved one public K-12 school (School A) in Hong Kong. Established in 1993, School A is an aided, co-ed primary school. School A has been incorporating STEM education since kicking off its first STEM Day in 2016. STEM education at the school is mainly implemented via the *Young Maker* Course, an "other learning experience". This course is designed by drawing topics and themes from the *General Studies* subject, which involves fundamental scientific and technological principles and the combination of information and technology elements. Students become the "makers" by engaging in the activities, applying related knowledge and skills, and stretching their potential in creativity. In the last three years, School A has also embraced a holistic approach. Most subjects in the school have included some STEM elements, although the level of integration varies.

School A was purposively selected because it is a pioneering and award-winning school in Hong Kong K-12 sector's STEM education. School A has employed several TPD models in conjunction since STEM education was introduced. School A's success in STEM education adoption and the varieties of TPD approaches gave the researchers the confidence in its uniqueness as an ideal case for examining necessary attributes and missing links of effective TPD for STEM education.

3.4 Data Analysis

Guided by our research questions, we utilised the inductive thematic analysis (Creswell 2014). We first transcribed the recorded audio files and organised the observation notes to make meaning out of them. We then annotated the data into themes, which involved a three-step process: (1) clustering the data into domains, (2) condensing the data into core ideas, and (3) cross-analysing to extract common themes across all participants. This process was conducted by two researchers independently. Upon completion, the team met again to discuss the domains and core ideas and extracted common themes. Such practices helped mitigate researcher misinterpretation or bias, which subsequently contributed to the validity and trustworthiness of the findings.

4 Findings and Discussions

School A took a mixed approach to TPD models as the school introduced STEM education. The cascade model, training model, partnership model and action research model were used in conjunction, and they are detailed in the following sub-sections.

4.1 Cascade Model of TPD Supports Laying the Groundwork for STEM Education

Principal Lily is the primary change agent of her school's successful STEM education adoption. She first learned about the concept in one of the public seminars conducted by a senior Hong Kong education leader in 2012. Over time, the principal has developed a strong vision for STEM education with the growing amount of information gathered. She believes that STEM education in her school should not only engage students' interest in the respective disciplines but also give them a competitive edge to become "future-ready". Such leadership vision became the driving force for implementing STEM education in her school (Cheng 2015). At the same time, she wants STEM education to be a key feature and the strength of her school in the eyes of parents and the public for sustaining school development. To achieve this vision, Principal Lily and her team have made many school improvement efforts such as renovating school buildings for STEM-friendliness, adding STEM Space and purchasing many STEM-related hardware and learning kits. She was also granted a Quality Education Fund (QEF) project for financing her schoolteachers to pilot innovative STEM education practices and develop school-based STEM learning resources. She served as the agent to create conditions to build professional learning communities to innovative pedagogies (Cheng 2017; Hords 1997).

At the time when Principal Lily and her senior management members had discussions on incorporating STEM education at the school curriculum level in 2013, the notion was a relatively new concept under consultation; it had not yet been formally included in Hong Kong education policy guidelines. The principal initiated STEM education strategically by prioritising it in the school development plan, a method that Cheng and Ko (2012) advocated. She noticed that most of the "good practices" of STEM education in Hong Kong education circles were heavily focused on teaching coding to secondary students, which, in her perception, was an important domain but not necessarily the ideal area of focus for primary school students. Despite concerns over what to be taught, particularly regarding the age-appropriateness of the content, Principal Lily and her team continued to explore this unknown territory, with STEM Day being launched at School A as an awareness-building and student-competition event on an annual basis. Meanwhile, Principal Lily decided to refer to new developments and promising practices of STEM education worldwide and work closely with EDB. The principal and selected school staff, therefore, actively attended STEM-related academic conferences and education fairs held in many countries and regions. These extensive exposures helped the principal learn about the "maker movement" and its practices in K-12 education globally. This notion was brought back to School A, and its curricular significance, curriculum organisation and implementation feasibilities were studied and then disseminated in a series of managerial meetings among school leaders and school-based TPD training events among teachers. These discussions resulted in the establishment of the "Young Maker" class as an "other learning experience" in the school-based curriculum.

Because the "Young Maker" class was one of the first of its kind in Hong Kong's K-12 education landscape, School A gained recognition from EDB and sister schools as being a pioneer in local STEM education. Such a success suggests that the cascade Model of TPD was highly effective in supporting School A's introduction of STEM education at its induction stage.

4.2 The Unavailing Training Model Prompts a Revisit to TPD Content

While the cascade model of TPD laid the groundwork for School A's STEM Education, teacher competencies for implementing the "Young Maker" class became a challenge. Because STEM education was a new, integrated curriculum area, most of the existing teachers did not receive any pre-service teacher training for it. As a result, there was not a critical mass of teachers who would be fully "capable" of teaching the "Young Maker" course. This situation echoes Chai (2019), Chai et al. (2019, 2020b) findings concerning the scarcity of well-trained teachers' to implement STEM. With respect to the key learning areas of the "Young Maker" class, the current General Studies, Science, Mathematics, and ICT subject teachers joined as the teaching team as an interim strategy. While the topics of the Maker lesson were extracted from General Studies subjects, teachers found that adding STEM elements to those topics was challenging. This is because General Studies teachers often lack a science or engineering background, not to mention familiarity with the interplay of these two disciplines. And those who did have such a background may not have known how to teach an integrated subject outside the bounds of their subject area expertise. Such reality reflects Honey et al.'s (2014) claim that limited curriculum integration skills could disable the implementation of STEM education.

To deal with this issue, all teachers in School A were encouraged to attend EDB's in-service training events on a regular basis.

In our interviews, Principal Lily stressed the importance of empowering teachers via training TPD and providing them with knowledge and skills about STEM pedagogy:

"As the principal, I am obligated to ensure our school's STEM education development is on the right track. I have personally joined many STEM education demonstrations and open classes in different countries and regions. But people tend to focus too much on what they use rather than how the class is taught. What they use are just consumable products. For any school introducing STEM education, purchasing related products is the easiest part. It is the teachers who can make good use of learning resources and design meaningful activities that will make a difference. For such, our teachers have to brush up on their skills through training."

Unfortunately, the keenness of teachers attending TPD training did not meet what the principal had hoped. It appeared that the teachers did not perceive any urgency to make a change in implementing the school-based STEM education, and the principal had to create an urgency for change (Kotter 1996). Despite her rather extraordinary individual efforts to promote TPD training, the principal acknowledged that some of the teachers still lacked the incentive and that she could not expect all the teachers to appreciate such an effort. She presumed that, for some teachers, teaching excellence was merely built on a deep-rooted tradition of student high academic achievement as a form of meritocracy. As the curricular demands and administrative tasks of teachers were already overwhelming, the mandate of extra TPD training could be challenging, particularly given the highly competitive education culture in Hong Kong.

However, our interviews revealed a different side of the story: the teachers viewed the training TPD as ineffective and unable to assist them fully for several pedagogical reasons.

First, as STEM education is an umbrella term covering highly diversified areas, the training events were often attended by teachers who possessed different disciplinary knowledge, different levels of teaching experience, different school-level teaching qualifications, and different interests in STEM education. This reality made it almost impossible for the training programme to cater to all of their needs, as what was reported in Cochran-Smith and Lytle's (1999) study. As a result, the training events were often introductory, and teachers could not have a deep dive into the topics they needed the most. One ICT teacher, who dismissed the content for its being "nothing technically new", said: *"To me, the content offered [in those events] is very basic. I think [this is] because most attendees do not have much technical background."* This is similar to the risk of dilution of quality discussed by Turner et al. (2017).

Second, probably because of the same reason, training had often focused on the technical perspective rather than a pedagogical perspective of teaching STEM education. While it was indeed beneficial for teachers without science and technology background to gain some technical knowledge on the STEM-related tools such as *Scratch* and *Micro Bit*, they may not develop an appreciation or adequate pedagogical understanding of how these tools can help them design meaningful learning activities and support the teaching of their STEM lessons. As one General Studies teacher mentioned: *"It is always good for us to learn these tools as we are non-technical people. But I am interested more in how I can make good use of these tools in my lessons, which the training seldom covered."*

Third, teachers found that the training events rarely connected with their current teaching contexts or classroom practices, and that some approaches contradicted their own teaching beliefs and guiding educational principles. For example, the Chinese and English language discipline teachers found it somewhat confusing when they were asked to infuse STEM elements, and the training TPD did not address their confusion. Although the teachers were told that the mentality of "trial and error" and "self-debugging" in STEM could be used in their lessons, they found such practices only had limited applications, such as in the writing classes, so they were not convinced that the method was an authentic means of teaching language acquisition. As one teacher explained: *"Language learning is not like the simple input and output of computers. It requires the development of meaning. I use online tools like Kahoot! in my lesson. But I'm not sure if these can be considered as STEM."* This finding echoes Kennedy's (2005) comments on the technicist view of teaching in the training model for its deficit of *"why"*.

4.3 Partnership Model of TPD Can Bring in New Knowledge, but Application Requires Sustainability and Teacher Collaboration

The partnership model was applied externally as means to extend teachers' horizons. School A has teamed with several sister schools by joining STEM education-related associations and alliances in Hong Kong and abroad. Principal Lily also sent teachers to mainland China or overseas in exchange programmes so that they could tap into others' expertise and experience. Meanwhile, because of her extended professional network, Principal Lily often invited university professors and industrial experts to School A to conduct sharing sessions such as STEM Talk and Maker Talk. These sessions not only allowed teachers to learn about the latest developments in the field, but also discussed the theoretical foundations of STEM education. Although a promising start to effective STEM education introduction, the outcomes of these TPD events did leave significant room for revisiting and refinement.

In our interviews, we found that some teachers were able to recognise STEM as the interplay between science, technology, engineering, and mathematics disciplines. Still, their conception of STEM education was often as four separate disciplines brought together under an umbrella term, instead of an integrated branch of study. This was also apparent in our analysis of the lesson plans and classroom observations, indicating there was a superficial acknowledgement of the connections between STEM disciplines. Document analysis of their lesson plans and lesson observations suggest that they simply combined knowledge or skills from several disciplines for addressing the specific topics and did not alter their pedagogical approaches when carrying out learning activities in the classroom. Though some teachers had learned that STEM education is not just about the acquisition of disciplinary knowledge but the development of metacognitive skills such as design thinking and computational thinking, the interviewees as a whole seemed to lack this more nuanced understanding. That lack of nuance may have inhibited the effective implementation of STEM pedagogy at School A.

This unfavourable situation may be partly due to the fact that School A's partnership model TPD events were often presented in a top-down, lecture-style format, and the fact that the events were sometimes held on a one-off basis and thus lacked a coherent progression of topics. A more structured and cohesive approach for the abovementioned seminars may have allowed School A's teachers to develop a more unified conception of STEM educational theories, enabling them to move beyond their familiarity with the current paradigm shift with STEM education and begin implementing more pedagogically sound methods in their classrooms.

Our classroom observations also revealed limited internal collaborative partnerships between STEM subject teachers at School A. Without such collaboration, teachers could not provide the kind of interdisciplinary expertise and support that students require to navigate complex STEM-related topics. Moreover, they experienced considerable difficulty in ensuring students complete the school curriculum at a reasonable pace and without unnecessary complications. For example, our interviews indicate that teachers at School A faced considerable challenges in designing topics that both satisfied the curriculum requirements while also being appropriate to the student's level. The latter is vital because STEM students have to complete advanced-level projects that require the integration of content knowledge from two or more disciplines. To ensure students are prepared for this, teachers need to continually reflect on STEM topic prerequisites and consult with each other on such issues as what content their students have covered and what are their strengths and weaknesses. Only collaboration on both fronts can ensure the lateral cohesion necessary for teachers to support the vertical progression of their students throughout the STEM education curriculum.

Teachers had a limited ability to achieve appropriacy in developing their topics also because of a lack of co-planning with their peers teaching other STEM disciplines. While School A has nurtured a culture of collaboration among teachers of individual disciplines, our observation suggests there is room for fostering greater cross-disciplinary collaboration. Their current professional dialogue tended to be *ad hoc*, 'just in time' and taking place in teachers' 'own time'. This placed limits on the opportunities for teachers to reflect on their classroom practice generally and, more specifically, in relation to STEM subject integration. TPD models like collaborative action research may help them to develop teacher collaboration and reflective practices (Mills 2007; Brown et al. 2015).

4.4 Action Research Model Can Encourage Innovative Practice of STEM Education, but Teachers Need Support

Principal Lily expressed in the interview that, despite her strong push on STEM education at the helm of the school, she did, in fact, believe that a bottom-up approach of TPD may have a more significant value for bringing change, and that such an approach could foster two-way interaction between teachers and the school:

"Instead of teachers waiting for the school to tell them what to do, new practices can be piloted or self-taught with the school's support...... Teachers' good practice can, in return, push the school to roll out more conducive policies towards STEM education."

With such beliefs, Principal Lily allocated resources and made supporting policies for teachers conducting collaborative action research so that they could continue to refine their knowledge and hone their instruction. For example, resources were made available for teachers to pilot new methods or new tools and share their findings in academic conferences. Principal Lily also shared STEM-related books and articles she came across to help teachers reflect on their teaching practices.

Despite such favourable conditions, action research was only practised by a few highly-motivated teachers in the school, and most teachers had not developed a passion for the method. Admittedly, building a culture of action research for TPD was difficult. As a cyclic and iterative method of systematic inquiry, action research is time intensive (Elliott 1976). And time is often a luxury for teachers considering their workload. At School A, teachers' lack of time, combined with their general lack of action research experience, posed a considerable challenge for expanding the scope of action research practice.

5 Conclusion and Implications

This study aims to identify the necessary attributes and missing links so that K-12 schools can incorporate these findings into their future TPD programmes. With the promising practices and lessons learned from our case study, the following guiding principles are given to promote effective TPD for STEM education.

First, TPD for STEM education should be school-based designed and contextualised with teachers' individual practices. They should be aligned with the goals of school-based STEM curricula, departmental resources provided, and teachers' curriculum organisation method. Extensive research evidence has pointed to the importance of leadership from principals (Cheng 2017; Cheng and Ko 2012; Fullan 2012)—they serve as a key driving force of effective TPD. Moreover, instead of a "one-size-fits-all" approach, TPD should be related to the various disciplinary focus among teachers.

Second, STEM education TPD needs an organic combination of TPD models. Mixed-use of TPD models can be an effective strategy for introducing STEM education, but sustaining good practice and moving to the next level may require more rigorous planning of how different models are best combined. It also needs to be dynamic and able to balance changing trajectories across the phases of a school's STEM education development.

Third, TPD for STEM education should help teachers build a shared vision. Before offering teachers *know-what* and *know-how*, they need to develop their own *know-why*. Our study findings are consistent with Thibaut et al. (2018) research result that teachers' attitudes are positively linked with their instructional practices. TPD, therefore, should help STEM teachers recognise the compelling and inherent opportunities of the paradigm to strengthen and support the teaching of STEM education, and, where possible, integrate STEM elements into the curriculum. Articulating the benefits of STEM education to students may create a sense of urgency for STEM implementation, and subsequently enable teachers to develop their own *know-why*.

Fourth, TPD must cater to teachers' various needs. As STEM education is interdisciplinary and involves teachers of various backgrounds, TPD should be more personalised so that teachers can better appreciate the vision and philosophies behind STEM education and be better equipped to apply their strengths in developing meaningful learning activities. This process will likely take time, as noted by Darling-Hammond et al. (2017). Therefore, TPD activities should be held over a sustained period, offering multiple opportunities for teachers to explore how their needs can be catered. Adult learning theory (Merriam 2018) may be applied for enhancing personalisation as various TPD models are being employed.

Fifth, STEM education TPD needs to focus on integration, and teachers' TPACK should be a key focus area for development (Chai et al. 2020a). While the interconnectedness between disciplinary knowledge of STEM is complex, dynamic, and contextual (Chai et al. 2019), teachers could lack a sense of a coherent and integrated STEM curriculum agenda. This missing link can hinder the effectiveness of their teaching. Therefore, it is critical to make the *what*, the *why* and the *how* of integration an explicit objective of TPD programmes. Only when teachers form clear answers to these questions for themselves, together with their TPACK, can they achieve good integration. Co-design can probably serve as an effective strategy for developing integration skills and uplifting teachers' TPACK (Wu et al. 2020).

Sixth, TPD for STEM should encourage a collaborative culture. Schools' successful adoption of STEM education requires concerted efforts from all stakeholders, and this includes collaboration between teachers. One strategy to achieve this is to adopt collaborative action research to exchange and craft STEM pedagogies and reflect on the issues together, thereby developing collective wisdom. Peer learning is another means of engendering teacher collaboration. This approach facilitates peer feedback or constructive criticism that can help scaffold teachers' action research and implementation of their reflection-based instructional practices. Peer feedback from like-minded colleagues helps teachers evolve their STEM education practices rather than provide prescribed solutions that do not fit their teaching beliefs, subject nature, and context.

We believe that, when these six guiding principles and their related strategies are considered, schools are more likely to move towards effective TPD for STEM Education.

References

Aslam, F., Adefila, A., Bagiya, Y.: STEM outreach activities: an approach to teachers' professional development. J. Educ. Teach. **44**(1), 58–70 (2018)

Avalos, B.: Teacher professional development in teaching and teacher education over ten years. Teach. Teach. Educ. **27**(1), 10–20 (2011)

Breiner, J.M., Harkness, S.S., Johnson, C.C., Koehler, C.M.: What is STEM? A discussion about conceptions of STEM in education and partnerships. Sch. Sci. Math. **112**(1), 3–11 (2012)

Brenneman, K., Lange, A., Nayfeld, I.: Integrating STEM into preschool education; designing a professional development model in diverse settings. Early Childhood Educ. J. **47**(1), 15–28 (2019)

Brown, B., Wilmot, D., Paton Ash, M.: Stories of change: the case of a foundation phase teacher professional development programme. S. Afr. J. Child. Educ. **5**(1), 191–209 (2015)

Bybee, R.W.: The Case for STEM Education: Challenges and Opportunities. NSTA Press, Arlington, VA (2013)

Central Policy Unit: Research report on smart city. Retrieved from The Government of The Hong Kong Special Administrative Region Website (2015). http://www.cpu.gov.hk/doc/en/research_reports/CPU%20research%20report%20-%20Smart%20City(en).pdf

Chai, C.S.: Teacher professional development for science, technology, engineering and mathematics (STEM) education: a review from the perspectives of technological pedagogical content (TPACK). Asia Pac. Educ. Res. **28**(1), 5–13 (2019)

Chai, C.S., Jong, M.S.-Y., Yin, H.-B., Chen, M., Zhou, W.: Validating and modelling teachers' technological pedagogical content knowledge for integrative science, technology, engineering and mathematics education. Educ. Technol. Soc. **22**(3), 61–73 (2019)

Chai, C.S., Rahmawati, Y., Jong, M.S.Y.: Indonesia science, mathematics, and engineering preservice teachers' experiences in STEM-TPACK design-based learning. Sustainability **12**(21), 1–14 (2020a)

Chai, C.S., Jong, M., Yan, Z.M.: Surveying China teachers' technological pedagogical STEM knowledge: a pilot validation of STEM-TPACK survey. Int. J. Mob. Learn. Organ. **14**(2), 203 (2020b)

Cheng, E.C.K.: Knowledge Management for School Education. Springer, London (2015)

Cheng, E.C.K.: Managing school-based professional development activities. Int. J. Educ. Manag. **31**(4), 445–454 (2017)

Cheng, E.C.K.: Successful Transposition of Lesson Study: A Knowledge Management Perspective. Springer, London (2018)

Cheng, E.C.K., Ko, P.Y.: Leadership strategies for creating a learning study community. Int. J. Mob. Learn. Organ. **9**(1), 161–180 (2012)

Cochran-Smith, M., Lytle, S.L.: Chapter 8: relationships of knowledge and practice: teacher learning in communities. Rev. Res. Educ. **24**(1), 249–305 (1999)

Creswell, J.W.: Qualitative Inquiry and Research Design: Choosing Among Five Approaches. Sage Publications, Thousand Oaks, CA (2007)

Creswell, J.W.: Research Design: Qualitative, Quantitative, and Mixed Methods Approaches, 4th edn. Sage Publications, Thousand Oaks, CA (2014)

Cross, N.: Design Thinking: Understanding How Designers Think and Work. Berg (2011)

Curriculum Development Council: Ongoing renewal of the school curriculum – focusing, deepening, and sustaining. Updating the technology education key learning area curriculum (primary 1 to secondary 6), consultation brief. Author, Hong Kong (2015). Retrieved from https://www.edb.gov.hk/attachment/en/curriculum-development/renewal/Brief_TEKLA_E.pdf

Darling-Hammond, L.: Powerful Teacher Education: Lessons from Exemplary Programs. Wiley, San Francisco (2006)

Darling-Hammond, L., Hyler, M.E., Gardner, M.: Effective Teacher Professional Development. Learning Policy Institute, Palo Alto, CA (2017)

Education Bureau of Government of HKSAR: Report on promotion of STEM education: Unleashing potential in innovation (2016). Retrieved from https://www.edb.gov.hk/attachment/en/curriculum-development/renewal/STEM%20Education%20Report_Eng.pdf

Elliott, J.: Developing hypotheses about classrooms from teachers'' practical constructs: an account of work of Ford teaching project. Interchange **7**(2), 2–22 (1976)

Freeman, S., et al.: Active learning increases student performance in science, engineering, and mathematics. Proc. Natl. Acad. Sci. **111**(23), 8410–8415 (2014)

Fullan, M.: Change Forces: Probing the Depths of Educational Reform. Routledge, London (2012)

Gaikhorst, L., Beishuizen, J.J., Zijlstra, B.J., Volman, M.L.: The sustainability of a teacher professional development programme for beginning urban teachers. Camb. J. Educ. **47**(1), 135–154 (2017)

Government of the Hong Kong Special Administrative Region: 2015 Policy Address—Uphold the Rule of Law, Seize the Opportunities, Make the Right Choices, Pursue Democracy, Boost the Economy, Improve People's Livelihood (2015). Retrieved from https://www.policyaddress.gov.hk/2015/eng/pdf/PA2015.pdf

Guskey, T.R.: Evaluating Professional Development. Corwin Press Inc., Thousands Oak, CA (2000)

Guskey, T.R.: Professional development and teacher change. Teachers and Teaching **8**(3), 381–391 (2002)

Han, S., Capraro, R., Capraro, M.M.: How science, technology, engineering, and mathematics (STEM) project-based learning (PBL) affects high, middle, and low achievers differently: The impact of student factors on achievement. Int. J. Sci. Math. Educ. **13**(5), 1089–1113 (2015)

Hardman, F.: Review: Teacher Development and Support Interventions. Save the Children Global Alliance, London (2011)

Harwell, S.H.: Teacher Professional Development: It's Not an Event, It's a Process. CORD, Waco, TX (2003)

Hayes, D.: Cascade training and teachers' professional development. ELT J. **54**(2), 135–145 (2000)

Honey, M., Pearson, G., Schweingruber, A.: STEM Integration in K-12 Education: Status, Prospects, and An Agenda for Research. National Academies Press, Washington, DC (2014)

Hooker, M.: Models and best practices in teacher professional development (2008)

Kelley, T.R., Knowles, J.G.: A conceptual framework for integrated STEM education. Int. J. STEM Educ. **3**(1), 1–11 (2016). https://doi.org/10.1186/s40594-016-0046-z

Kemmis, S.: Action research. In: Keeves, J.P. (ed.) Educational Research Methodology and Measurement: An International Handbook, pp. 237–253. Pergamon, Oxford (1988)

Kennedy, A.: Models of continuing professional development: a framework for analysis. J. In-Serv. Educ. **31**(2), 235–250 (2005)

Koehler, M.J., Mishra, P., Cain, W.: What is technological pedagogical content knowledge (TPACK)? J. Educ. **193**(3), 13–19 (2013)

Kotter, J.P.: Leading Change. Harvard Business School Press, Boston (1996)

Lamberg, T., Trzynadlowski, N.: How STEM academy teachers conceptualise and implement STEM education. J. Res. STEM Educ. **1**(1), 45–58 (2015)

Laurillard, D., Kennedy, E., Wang, T.: How could digital learning at scale address the issue of equity in education? In: Lim, C.P., Tinio, V.L. (eds.) Learning at Scale for the Global South. Foundation for Information Technology Education and Development, Quezon City, Philippines (2018)

Laurillard, D., Kennedy, E.: Digital multiplier model for teacher professional development at scale. Foundation for Information Technology Education and Development, Quezon City, Philippines (2019)

Margot, K.C., Kettler, T.: Teachers' perception of STEM integration and education: a systematic literature review. Int. J. STEM Educ. **6**(1), 1–16 (2019). https://doi.org/10.1186/s40594-018-0151-2

Marshall, C., Rossman, G.B.: Designing Qualitative Research, 5th edn. Sage Publications, Thousand Oaks, CA (2011)

Merriam, S.B.: Qualitative Research: A Guide to Design and Implementation. Jossey-Bass, San Francisco (2009)

Merriam, S.B.: Adult learning theory: evolution and future directions. In: Illeris, K. (ed.) Contemporary Theories of Learning, pp. 83–96. Routledge, London (2018)

Merrill, C., Daugherty, J.: The future of TE masters degrees: STEM. Paper presented at the meeting of the International Technology Education Association, Louisville, KY (2009)

Mills, G.E.: Action Research: A Guide for the Teacher Researcher, 3rd edn. Merrill Prentice Hall, New York (2007)

Moore, T.J., Johnson, C.C., Peters-Burton, E.E., Guzey, S.S.: The need for a STEM road map. In: Johnson, C.C., Peters-Burton, E.E., Moore, T.J. (eds.) STEM Road Map: A Framework for Integrated STEM Education, pp. 3–12. Routledge, New York (2015)

National Academies of Sciences, Engineering, and Medicine: Graduate STEM Education for the 21st Century. National Academies Press, Washington, DC (2018)

OECD: Creating Effective Teaching and Learning Environments: First Results from TALIS. OECD Publishing, Paris (2009)

OECD: OECD Science, Technology and Innovation Outlook 2018: Adapting to Technological and Societal Disruption. OECD Publishing, Paris (2018)

Ono, Y., Ferreira, J.: A case study of continuing teacher professional development through lesson study in South Africa. S. Afr. J. Educ. **30**(1), 59–74 (2010)

Opfer, V.D., Pedder, D.: Conceptualising teacher professional learning. Rev. Educ. Res. **81**(3), 376–407 (2011)

Pearson, G.: National academies piece on integrated STEM. J. Educ. Res. **110**(3), 224–226 (2017)

Pickering, T.A., Yuen, T.T., Wang, T.: STEM conversations in social media: implications on STEM education. In: Proceedings of IEEE International Conference on Teaching, Assessment and Learning for Engineering 2016 (TALE 2016), pp. 296–302. IEEE

Razzouk, R., Shute, V.: What is design thinking and why is it important? Rev. Educ. Res. **82**(3), 330–348 (2012)

Sanders, M.: Integrative STEM education: primer. Technol. Teach. **68**(4), 20–26 (2009)

Senge, P.M.: Schools That Learn: A Fifth Discipline Fieldbook for Educators, Parents, and Everyone Who Cares About Education, 2nd edn. Nicholas Brealey Publishing, New York (2012)

Schon, D.A.: The Reflective Practitioner: How Professional Think in Action. Temple Smith, London (1983)

Schwab, K.: The Fourth Industrial Revolution. Crown Business, New York (2017)

Strimel, G.J., Bartholomew, S.R., Kim, E., Zhang, L.: An investigation of engineering design cognition and achievement in primary school. J. STEM Educ. Res. **1**(1–2), 173–201 (2018)

Strimel, G.J., Kim, E., Grubbs, M.E., Huffman, T.J.: A meta-synthesis of primary and secondary student design cognition research. Int. J. Technol. Des. Educ. **30**(2), 243–274 (2019). https://doi.org/10.1007/s10798-019-09505-9

Svendsen, B.: Teachers' experience from a school-based collaborative teacher professional development programme: reported impact on professional development. Teach. Dev. **20**(3), 313–328 (2016)

Thibaut, L., Knipprath, H., Dehaene, W., Depaepe, F.: The influence of teachers' attitudes and school context on instructional practices in integrated STEM education. Teach. Teach. Educ. **71**, 190–205 (2018)

Turner, F., Brownhill, S., Wilson, E.: The transfer of content knowledge in a cascade model of professional development. Teach. Dev. **21**(2), 175–191 (2017)

Tytler, R., Appelbaum, P., Swanson, D.: Subject matters of science, technology, mathematics and engineering. In: He, M.F., Schultz, B.D., Schubert, W.H. (eds.) The SAGE Guide to Curriculum in Education, pp. 27–35. SAGE Publication, Newbury Park, CA (2015)

Ufnar, J.A., Shepherd, V.L.: The scientist in the classroom partnership program: an innovative teacher professional development model. Prof. Dev. Educ. **45**(4), 642–658 (2019)

Wang, T., Cheng, E.C.K.: An investigation of barriers to Hong Kong K-12 schools incorporating artificial intelligence in education. Comput. Educ. Artif. Intell. **2**, 1–11 (2021)

Wang, T., Cheng, E.C.K.: Towards a tripartite research agenda: a scoping review of artificial intelligence in education research. In: Cheng, E.C.K., Koul, R.B., Wang, T., Yu, X. (eds.) Artificial Intelligence in Education: Emerging Technologies, Models and Applications, pp. 3–24. SpringerNature, Singapore (2022)

Wing, J.M.: Computational thinking. Commun. ACM **49**(3), 33–35 (2006)

Wu, S., Peel, A., Bain, C., Anton, G., Horn, M., Wilensky, U.: Workshops and co-design can help teachers integrate computational thinking into their k-12 stem classes. In: Proceedings of International Conference on Computational Thinking Education 2020, pp. 63–68. The Education University of Hong Kong, Hong Kong (2020)

Yin, R.K.: Case Study Research: Design and Methods, 3rd edn. Sage Publications, Thousand Oaks, CA (2003)

Zeichner, K.: Teacher research as professional development for P-12 educators in the USA. Educ. Action Res. **2**(2), 301–326 (2003)

Identification and Hierarchical Analysis of Risk Factors in Primary and Secondary Schools: A Novel GT-DEMATEL-ISM Approach

Shuzhen Luo[1], Qian Wang[1(✉)], and Peirong Qi[2]

[1] School of Safety and Emergency Management Engineering, Taiyuan University of Technology, Taiyuan 030000, Shanxi, China
lszlbo@163.com, wenyq2001@163.com

[2] Shuangta North Road Primary School, Yingze District, Taiyuan 030000, Shanxi, China
qpr2011@126.com

Abstract. As an important part of public safety throughout the world, primary and secondary school safety has been faced with increasingly diverse and complex risks. Literature mentions different methods to identify and assess safety risks in primary and secondary schools. However, none of these methods help in defining a complete scope. Differently, this study aims to clarify the relationship between risk factors and their impact on school safety by integrating qualitative and quantitative research methods. Firstly, adopting grounded theory as the methodology, data collection entailed from two sources: Semi-structured interviews with experienced primary and secondary school staff (n = 20) across diverse schools, 231 cases of school safety accidents in primary and secondary schools in China from 2011 to 2021. The results show that 76 initial concepts, 19 initial categories, and 8 main categories are obtained through the coding step of grounded theory. Then, the 19 initial categories are defined as 19 risk factors in primary and secondary schools. We employed the combination of Decision-making Trial and Evaluation Laboratory (DEMATEL) and Interpretive Structural Modeling (ISM) techniques for understanding the hierarchal and contextual relationship structure among the 19 risk factors. This present novel model helps the policy and decision-makers to find out a mutual relationship and interlinking with school safety.

Keywords: Risk identification · Grounded theory · DEMATEL · ISM · Primary and secondary schools

1 Introduction

As an important part of public safety, the risks faced by primary and secondary school campuses are increasingly diverse and complex, and because most of the primary and secondary school campus subjects are minors under the age of 18, their physical and psychological development is insufficient, they lack social experience, and their ability to recognize risks, avoid risks and defend themselves is weak. According to the China Emergency Education and Campus Safety Development Report 2021, the distribution of school segments in which campus safety incidents occur is particularly notable in primary and secondary schools [1]. Therefore, it is important to study the identification and evolutionary levels of risk factors in primary and secondary school campuses and explore the process of incident development to ensure normal order and social stability in primary and secondary schools.

Risks in education are the product of the interaction between school and social risks and are characterized by diversity, complexity, and sociality, and their formation is cumulative, coupled, and diffuse [2]. Therefore, a systematic approach is needed to identify and assess risks in schools. Some studies have identified campus risks through questionnaires [3] and Fault Tree Analysis [4], while others have analyzed the relationships between factors using accident causation analysis [5] and Analytic Hierarchy Process which could be used to construct an evaluation index system [4, 6–9], and others have assessed the significance of the factors by Multiple Criteria Decision Making [10], Preventive Assessment [11], The Entropy method[12], Hierarchical Grey Theory and Analytic Hierarchy Process [13].

Firstly, the identification of risk factors is often based on personal experience, which is highly subjective, so the risk factors in primary and secondary schools are not systematically identified, and the generality and connotation of the identified factors are not uniform. Thirdly, the assessment of risk factors does not reflect the key indicators such as the strength of the impact of each risk factor due to the assessment method. To address these issues, a more explanatory, inductive, and exploratory approach is needed to identify the representations and commonalities of risk factors systematically and comprehensively on campus from objective phenomena.

Grounded Theory (GT) is a systematic and normative qualitative research method [14], which can effectively avoid the interference of pre-determined theories in the understanding of phenomena, and has been widely used in the study of motivation formation mechanisms and identification of safety risk factors in the fields of public health [15], construction [16] and special equipment [17]. The combined use of Decision Making Trial and Evaluation Laboratory (DEMATEL) and Interpretative Structural Modeling (ISM) can overcome the disadvantages of their separate use for the study of complex interfactor It has been widely used in the fields of waterway transportation [18], COVID-19 [19], software project [20] and subways [21] to study the complex relationships between factors and their influence on accidents.

Therefore, this paper proposes to use the advantages of objective induction of rooting theory and systematic modeling of DEMATEL-ISM method to reduce the influence of subjective factors on the construction of indicators and relationship analysis while retaining the logic of subjective reasoning to identify the hierarchical structure and exploration of the mechanism of action of key factors at all levels in primary and secondary school campuses, to provide reference and reference for school safety management.

2 Identification of Risk Factors for School Safety

2.1 Obtaining Data

Semi-structured Interviews. Following the principle of theoretical sampling of grounded theory [22]. 30 staff of primary and secondary schools were selected for in-depth interviews, including 5 primary and secondary school headmasters, 9 primary and secondary school classroom teachers, and 16 subject teachers in Shanxi Province. Interviews were conducted through offline field visits or online QQ and WeChat voice calls and were recorded using mobile phone recording software. An outline of the interview was sent to the interviewees before the interview to ensure that it opened up their minds. The interviews focused on the topic of campus safety risks, guiding the interviewees to recall incidents of safety risks on campus and the arrangements of the management-related system, introduce activities or incidents they have personally experienced in managing or governing campus safety risks, express their feedback and evaluation of the current campus risk management system or governance mechanism, and provide suggestions for improving the campus safety risk management system in the future. A total of 187 valid interview statements were obtained from the interview.

Collection of School Accident Cases. A total of 231 cases of school safety accidents in primary and secondary schools from 2011 to 2021 were collected for analysis. The cases mainly focused on accidents reported by the online media, with the main sources being the official website of the Ministry of Education of the People's Republic of China, People's Daily Online, Guangming.com, CCTV News, and other authoritative media reports. For the sake of coding, only those cases where the causes and processes of accidents were made public were selected.

2.2 Analysis of Coding

To construct a comprehensive campus risk factor identification system based on the rooting theory, it is necessary to study the interview texts and sample literature already collected, follow the steps of open coding, spindle coding, and selective coding, analyze and review the generalization and expression of campus risks in the textual materials, and achieve coding by statistically integrating the sample textual materials with the help of the coding function of the qualitative research software NVivo11, to achieve an efficient and in-depth analysis of a large amount of literature.

Open Coding. Open coding is one of the three basic analytical processes of grounded theory analysis. It requires a sentence-by-sentence, line-by-line, and word-by-word analysis of raw data materials such as interview transcripts and accident cases. Phrases that occur more frequently are first summarized as the initial set of concept candidates for open coding and initially coded, then the phrases in this candidate set are deeply condensed, which in turn leads to categories that can summarize similar concepts, then based on the initial set of concepts, the individual initial concepts need to be continuously compared over and over again, and in the process of comparison, the individual concepts are repeatedly modified and validated. Considering the limitation of space, the presentation of the conceptualization and categorization of the original text is omitted, resulting in 76 initial categories and 19 categories (columns 3 and 2 of Table 1).

Table 1. Results of axial coding of risk factors in primary and secondary schools.

Main categories	Initial categories	Initial concept
Natural disaster risk	S_1(Natural hazard causative factors)	Floods, lightning, earthquakes, landslides, landslides, mudslides, ground collapses, typhoons
	S_2(Vulnerability of hazard-bearing bodies)	Topography, landform, water system, surrounding layout, building resilience
	S_3(Disaster prevention and mitigation capacity)	Disaster prevention and mitigation awareness, emergency response capacity, early warning power, resilience
School facility risk	S_4(Factors of facilities)	Construction of facilities not meeting safety standards, damage to facilities
	S_5(Factors of using facilities)	Unsafe use of facilities by persons, failure use of facilities by persons
	S_6(Factors of facility management)	Irregularities in the management of facilities and equipment, failure to repair safety hazards in facilities timely
Accidental injury risk	S_7(Exercise and activity factors)	Sports activities, students playing with each other (lack of awareness of safety precautions)
	S_8(Congestion in school)	Accidents caused by congestion in buildings, gates, sports grounds, etc. on campus
	S_9(Traffic accidents in school)	Traffic accidents occurring in or around the school, or on a school bus
Public health risk	S_{10}(Public health causative factors)	Bacterial food poisoning, fungal toxin poisoning, plant food poisoning, chemical food poisoning
		Influenza, chickenpox, rubella, mumps, hand, foot and mouth disease, norovirus, novel coronavirus
		Air, water and land pollution

(*continued*)

Table 1. (*continued*)

Main categories	Initial categories	Initial concept
	S_{11}(Public health management vulnerability)	Food, health and epidemic, environmental and public health emergency management system
School emergency security risk	S_{12}(Law and order threatening factors)	Dangerous off-school people, students in trouble
	S_{13}(Security Vulnerability)	Access control system, surveillance system, lighting, security response capability, police and school linkage capability
School bullying risk	S_{14}(School Bullying Risk Derived Environment)	Failure of the education administration to restrain bullying, failure of the school to detect or deal with bullying, and failure of the family to communicate in a timely manner
	S_{15}(School Bullying Vulnerability)	Bullying Exposure, Bullying Student Tolerance, Bullying Student Tolerance Resilience
Individual health risk	S_{16}(Individual health risk triggers)	Scenario-based triggers (physical education classes, physical tests, sports, military training)
	S_{17}(Individual Health Vulnerability)	Stressful triggers (school completion, excessive family or self-imposed demands, family tensions)
School group risk	S_{18}(Rights-based school group factors)	Large-scale strikes, demonstrations or other disruptive activities carried out by students or staff to defend their rights
	S_{19}(Anger-venting school group factors)	Large-scale group activities organized by students or their parents due to dissatisfaction with the school's handling of various incidents

Axial Coding. Based on the concepts and categories inducted by the open coding, the main and sub-categories were developed by surrounding a category, using causal, situational, time-series, and similarity relationships to achieve connections between the categories [23] and to explore the logical relationships between the categories. With the help of the standard paradigm of axial coding, the interrelationships between the 19 categories were analyzed, resulting in eight main categories: Natural disaster risk, school facility risk, accidental injury risk, public health risk, school emergency security risk, school bullying risk, individual health risk, and school group risk (column 1 of Table 1).

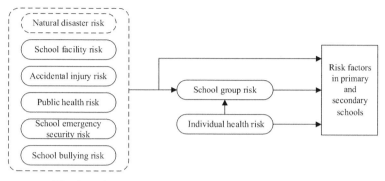

Fig. 1. A framework of factors formed by selective coding.

Selective Coding. Selective coding refers to a systematic and comprehensive analysis of the primary categories, which has been summarized so that the relationship between core and secondary categories is made clear by the causal model. Ultimately, this paper describes the relationship of each factor by focusing on the storyline of "school risk", which could be used as a core category to develop a causal analysis: Natural disaster risk, school facility risk, accidental injury risk, public health risk, school emergency security risk, school bullying risk, and individual health risk may further lead to risk addition, Fig. 1 shows the framework of factors influencing risk in primary and secondary schools based o grounded theory.

Saturation Test. A saturation test is required to ensure the reliability and completeness of the theoretical architecture construction. The theoretical structure was considered saturated if the categories obtained from the re-collected data material were included in the existing categories, no new categories emerged, and no new logical or causal relationships emerged between the relevant categories [24]. By coding the 20% of the sample set aside (interview texts and accident cases) in line with the previous section, no new categories emerged, which indicates that the theoretical structure passed the theoretical saturation test and that data collection and analysis could cease.

3 Model Construction of DEMATEL-ISM

3.1 Data Collection

To analyze the relationship between the factors corresponding to the 19 initial categories of risk in primary and secondary schools (for ease of presentation, the term "factors" is used below to denote "factors corresponding to the categories identified by the grounded theory"). In this paper, a relationship questionnaire was designed using the

Delphi method, 5 experts engaged in school safety research in universities and 15 people in charge of school safety management in primary and secondary schools were invited to score the strength of the relationship between the above risks (S_1 to S_{19}) based on a 0 to 3 scoring method: 0 (no impact), 1 (low impact), 2 (high impact) and 3 (very high impact) [24]. A total of 20 questionnaires were distributed and 20 valid questionnaires were returned, with a return rate of 100%.

3.2 DEMATEL Analysis

Step1. Calculate the Normalized Direct Influence Matrix B. $A\left(A = [a_{ij}]_{n \times n}\right)$ is the initial direct impact matrix A of risk factors in primary and secondary schools was obtained by pooling and averaging the data from 20 questionnaires, in which a_{ij} is denoted as the degree of influence of risk factor S_i to S_j in primary and secondary schools, $a_{ij} = 0$ when $i = j (i, j = 1, 2, \cdots n)$. To normalize a_{ij} to, the normalized direct impact matrix B could be obtained through the following Eq. (1):

$$B = \frac{A}{\max\limits_{1 \leq i \leq n} \sum_{j=1}^{n} a_{ij}} \quad (1)$$

where max denotes the maximum value of the sum of all row elements in the initial direct influence matrix A.

Step 2. Calculate the Comprehensive Influence Matrix T. The comprehensive influence matrix T could be obtained by coupling direct and indirect interactions between factors through the following Eq. (2), and the values are shown in Table 2.

$$T = B(I - B)^{-1} \quad (2)$$

Here I refer to the unit matrix, indicating the impact of the factor on itself.

Step 3. Determining the Influencing Degree, Influenced Degree, Centrality Degree and Causality Degree. The summation of rows of the elements were D_i in the comprehensive influence matrix T, which indicates the total given both direct and indirect effects. And the summation of columns of the elements were R_i in the comprehensive influence matrix T, which indicates the total received both direct and indirect effects. Then, the (D + R) represents the "centrality degree". Furthermore, "causality degree" is represented by (D-R). All those values are shown in Table 3.

According to Table 3, with the centrality degree as the horizontal coordinate and the cause degree as the vertical coordinate, a causality diagram is drawn as shown in Fig. 2. Among them, the influencing factors with vertical coordinates greater than 0 are causal factors, which are factors that directly affect the risk of primary and secondary schools; The influencing factors with vertical coordinates less than 0 are consequential factors, which indirectly affect the risk of primary and secondary schools due to the influence of causal factors.

3.3 ISM Analysis

Step 1. Calculate Reachability Matrix K. Since the ISM method uses binary numbers (0, 1) to represent the presence or absence of influence relationships between factors, it

Identification and Hierarchical Analysis of Risk   27

Table 2. The comprehensive influence matrix T.

Factors	S1	S2	S3	S4	S5	S6	S7	S8	S9	S10	S11	S12	S13	S14	S15	S16	S17	S18	S19
S1	0.024	0.127	0.110	0.124	0.084	0.072	0.110	0.168	0.111	0.077	0.052	0.121	0.107	0.065	0.061	0.196	0.115	0.124	0.164
S2	0.102	0.082	0.157	0.135	0.132	0.121	0.236	0.243	0.189	0.068	0.057	0.192	0.214	0.115	0.112	0.199	0.141	0.142	0.249
S3	0.066	0.181	0.095	0.204	0.233	0.212	0.272	0.271	0.190	0.072	0.133	0.146	0.213	0.139	0.123	0.209	0.189	0.205	0.346
S4	0.071	0.163	0.101	0.110	0.215	0.195	0.220	0.226	0.124	0.052	0.038	0.147	0.213	0.103	0.098	0.251	0.156	0.227	0.301
S5	0.034	0.122	0.150	0.188	0.132	0.197	0.218	0.228	0.120	0.051	0.041	0.122	0.217	0.125	0.100	0.185	0.194	0.202	0.292
S6	0.060	0.177	0.118	0.209	0.231	0.119	0.258	0.226	0.132	0.053	0.040	0.164	0.165	0.157	0.141	0.241	0.165	0.243	0.315
S7	0.043	0.117	0.080	0.152	0.223	0.181	0.180	0.249	0.181	0.066	0.046	0.162	0.154	0.187	0.196	0.281	0.243	0.171	0.314
S8	0.051	0.135	0.114	0.179	0.226	0.204	0.240	0.166	0.164	0.097	0.052	0.193	0.165	0.125	0.123	0.261	0.234	0.186	0.302
S9	0.009	0.027	0.028	0.039	0.048	0.040	0.069	0.131	0.060	0.027	0.017	0.152	0.084	0.049	0.046	0.165	0.059	0.067	0.176
S10	0.005	0.015	0.022	0.028	0.037	0.031	0.050	0.040	0.050	0.028	0.100	0.035	0.053	0.033	0.033	0.127	0.111	0.152	0.175
S11	0.016	0.046	0.115	0.065	0.081	0.118	0.126	0.098	0.090	0.129	0.037	0.079	0.095	0.080	0.066	0.198	0.145	0.181	0.233
S12	0.021	0.076	0.089	0.079	0.100	0.084	0.205	0.192	0.193	0.091	0.053	0.114	0.156	0.174	0.160	0.239	0.149	0.193	0.259
S13	0.034	0.099	0.138	0.117	0.135	0.127	0.236	0.236	0.170	0.084	0.058	0.219	0.142	0.184	0.195	0.238	0.160	0.208	0.320
S14	0.025	0.072	0.066	0.133	0.167	0.152	0.255	0.221	0.161	0.059	0.038	0.217	0.200	0.121	0.207	0.280	0.235	0.212	0.327
S15	0.018	0.051	0.049	0.078	0.102	0.082	0.217	0.184	0.130	0.050	0.032	0.179	0.169	0.186	0.101	0.244	0.200	0.170	0.279
S16	0.021	0.061	0.058	0.133	0.168	0.118	0.226	0.164	0.163	0.077	0.060	0.174	0.152	0.155	0.135	0.162	0.210	0.214	0.292
S17	0.023	0.065	0.062	0.141	0.205	0.127	0.242	0.185	0.199	0.130	0.072	0.133	0.146	0.151	0.183	0.265	0.141	0.165	0.313
S18	0.004	0.012	0.016	0.014	0.016	0.015	0.028	0.031	0.076	0.010	0.007	0.031	0.121	0.021	0.022	0.033	0.019	0.029	0.107
S19	0.004	0.012	0.016	0.015	0.017	0.016	0.029	0.034	0.097	0.011	0.007	0.034	0.124	0.023	0.023	0.037	0.021	0.088	0.050

Table 3. Degree of influence.

Factors	D	Rating of D	R	Rating of R	D + R	Rating of D + R	D-R	Rating of D-R
S1	2.012	14	0.632	19	2.644	18	1.38	3
S2	2.885	10	1.639	15	4.524	14	1.246	4
S3	3.500	1	1.582	16	5.082	11	1.919	1
S4	3.011	7	2.139	13	5.151	10	0.872	7
S5	2.918	9	2.554	10	5.472	6	0.363	9
S6	3.216	4	1.834	11	5.422	12	1.834	2
S7	3.228	2	3.415	3	6.644	1	0.187	15
S8	3.217	3	3.296	4	6.512	3	0.079	13
S9	1.293	16	2.602	9	3.895	15	1.309	17
S10	1.125	17	1.233	17	2.358	19	0.108	14
S11	2.000	15	0.942	18	2.942	17	1.058	5
S12	2.628	12	2.613	8	5.241	9	0.015	12
S13	3.100	6	2.889	6	5.989	4	0.211	10
S14	3.148	5	2.194	12	5.341	8	0.954	6
S15	2.518	13	2.124	14	4.642	13	0.394	8
S16	2.744	11	3.812	2	6.556	2	1.068	16
S17	2.946	8	2.887	7	5.833	5	0.059	11
S18	0.613	19	3.177	5	3.790	16	2.565	18
S19	0.658	18	4.814	1	5.472	7	4.156	19

is necessary to find the threshold λ to eliminate the weak relationships between factors and to simplify the system structure in combination DEMATEL and ISM. The mean α of all elements in the overall influence matrix T and the standard deviation β of all elements are solved by MATLAB software to obtain: $\alpha = 0.1295$, $\beta = 0.0781$, $\lambda = \alpha + \beta = 0.2676$. The transformation from the overall influence matrix to the adjacency matrix L is shown in Eq. (3) and (4).

$$L_{ij} = \begin{cases} 1 h_{ij} \geq \lambda \\ 0 h_{ij} < \lambda \end{cases} \tag{3}$$

$$L = [l_{ij}]_{n \times n}, (i, j = 1, 2, \cdots, n) \tag{4}$$

The reachable matrix K is obtained by adding the adjacency matrix L to the unit matrix I and performing a redundancy removal process.

Step 2. Levels Partitioning. The reachable set $R(S_i)$ and the prior set $Q(S_i)$ and their intersection are obtained by hierarchically processing the reachable matrix K using the

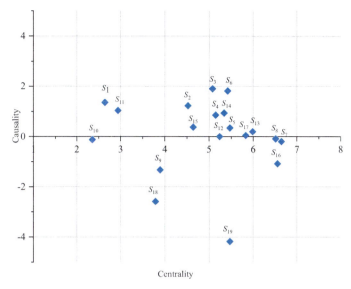

Fig. 2. Cause and effect diagram for elements.

following equation.

$$R(S_i) = \{S_j | S_j \in S, k_{ij} = 1\} (i = 1, 2, \cdots, n) \quad (5)$$

$$Q(S_i) = \{S_j | S_j \in S, k_{ji} = 1\} (i = 1, 2, \cdots, n) \quad (6)$$

$$P(S_i) = R(S_i) \cap Q(S_i) (i = 1, 2, \cdots, n) \quad (7)$$

When $P(S_i) = R(S_i)$, taking $P(S_i)$ as the highest level of influence factor, and remove the rows and columns of the queue of this influence factor from the reachable matrix K. Continue the above steps for iterative division until the levels of all factors are determined, until all factors are crossed out when they stop, and finally divide the system factors into 6 levels, and construct a comprehensive DEMATEL-ISM model of risk factors in primary and secondary schools according to the summary table of level decomposition and the centrality of each type of factor (Fig. 3).

The presence of a directional arrow between factors in the Fig. 3 implies that there is a causal relationship between the factors. If there are bidirectional arrows between the factors, the factors are said to be connected in a circuit, and they are strongly connected, indicating that the group of factors are causally related to each other.

3.4 Discussions

In the DEMATEL analysis, S_3 (Disaster prevention and mitigation capacity) and S_6 (Factors of facility management) are classified as causal factors with the highest degree of causality. In the ISM analysis, those two are also root cause predisposing factors and are located at the starting point of the ISM model. Both methods treat S_3 (Disaster prevention and mitigation capacity) and S_6 (Factors of facility management) as the most influential factors, being the root causes of the outcome factors or superordinate factors. Similarly, the centrality of S_7 (Exercise and activity factors) and S_{16} (Individual health risk triggers) are ranked in the top 2, indicating that they are important triggers that are closely linked to other factors; and these 2 factors can also be seen in the ISM model with multiple inflow and outflow arrows at key nodes of the model. Therefore, the analytical results obtained by the DEMATEL and the ISM are consistent and indicate that the research framework of the DEMATEL method and the ISM method is reliable.

As seen in Fig. 3, the factors influencing risk in primary and secondary schools form a hierarchical structure with a top-down hierarchy of 6 levels and 3 stages, with Level-1 factors forming the proximate causal stage, the middle Level-2, Level-3 and Level-4 factors forming the transitional causal stage, and the lower Level-5 and Level-6 factors forming the essential causal stage. The cause attribute (affected attribute) is stronger the further down the hierarchy, and the outcome attribute (affected attribute) is stronger the further up the hierarchy.

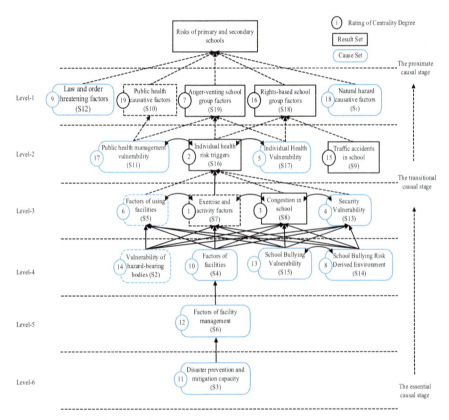

Fig. 3. DEMATEL-ISM comprehensive model of risk factors.

The Proximate Causal Stage. All factors within Level-1 of the multi-layered progressive explanatory structure model, including S_1(Natural hazard causative factors), S_{10}(Public health causative factors), S_{12}(Law and order threatening factors), S_{18}(Rights-based school group factors) and S_{19}(Anger-venting school group factors). As the uppermost factor in the model, it is only influenced by other factors but not other factors, indicating that factors at this stage are the most direct factors affecting risk in primary and secondary schools, and to achieve effective regulation of risk in primary and secondary schools, which can start with factors in the proximate causative order. It should be noted that some factors in this stage are susceptible to the influence of factors in other stages, so when controlling factors in this stage, it is necessary to control or cut off the links between them and other factors in a timely manner.

The Transition Causation Stage. All factors within levels 2, 3 and 4 of the multi-layered stepwise explanatory structure model, including S_5(Factors of using facilities), S_7 (Exercise and activity factors), S_8 (Congestion in school), S_9 (Traffic accidents in school), S_{11} (Public health management vulnerability), S_{13} (Security Vulnerability), S_{16} (Individual health risk triggers), S_{17} (Individual Health Vulnerability), S_{14} (School Bullying Risk Derived Environment), S_4 (Factors of facilities), S_2 (Vulnerability of hazard-bearing bodies) and S_{15} (School Bullying Vulnerability). Located in the middle segment of the overall model, they can both be influenced by and influence other risk factors. Transitional causal order factors can be influencing factors in their own right and assume a transitional (propagating) influence in the system. Therefore, it is necessary to control the factors of the transitional causal order, especially the factors with high centrality: S_7 (Exercise and activity factors) and S_{16} (Individual health risk triggers). These two factors link most of the factors in the overall system and deserve focused attention.

The Essential Causal Stage. All factors within Level- 5 and Level-6 of the multi-layered progressive explanatory structure model, including S_3 (Disaster prevention and mitigation capacity) and S_6 (Factors of facility management), mainly emitting directed arrows, are fundamental factors and the starting point of the DEMATEL-ISM integrated model of safety risks in primary and secondary schools, which will continue to influence other factors in the system in the long term, and should be focused on when identifying risks in primary and secondary school campuses, and should be considered when implementing Risk prevention and control efforts should be prioritized.

4 Conclusions

This paper presents an analysis on safety risks in primary and secondary schools by systematically combining GT, DEMATEL and ISM, which extend the existing research and understanding of the scope definition elements of safety risks in primary and secondary schools, and a framework has been proposed to analyze the relationship structure of it. Then main conclusions include:

Firstly, this paper systematically identifies the 3-level coding of predisposing factors for school risk in primary and secondary schools using the grounded theory, and obtains a total of 76 initial concept, 19 initial categories and 8 main categories. The predisposing factors for school risk in primary and secondary schools involve 19 factors in eight dimensions: Natural disaster risk, school facility risk, accidental injury risk, public health risk, school emergency security risk, school bullying risk, individual health risk, and school group risk.

Secondly, A comprehensive DEMATEL-ISM model of critical risks in primary and secondary schools was obtained, which verifies interactions among various factors. There are differences in the mechanisms and importance of different predisposing factors for school safety in primary and secondary schools. The DEMATEL analysis output gave a detailed depiction of cause-effect relationship of the factors exerted on the other. And the ISM can transform the fragmented and irregular elements of a complex system into a structured model with a clear hierarchy.

References

1. GAO Shan. Report on Chinese Emergency Education and Campus Safety Development. China Social Sciences Press.18+23 (2020)
2. NI Juan. Educational risks: A new perspective of educational research from the perspective of overall security Shanghai Research on Education.25–30 (2019)
3. YE Zhiqing. Research on the security governance of primary schools in Shenzhen's urban-rural fringe. Hunan University .25–26 (2018)
4. Shang, L.I.U., Chun, L.I.U., Xianqin, W.U., et al.: Research on the causes of stampede accidents in primary and secondary schools based on fault tree-analytic hierarchy process. Saf. Environ. Eng. **25**(04), 139–145 (2018)
5. ZHANG Jinnan. Research on campus safety accidents from the perspective of causal theory. Shandong University,**27** (2020)
6. Dessy Seri Wahyuni: Ketut Agustini, and Gede Ariadi, "An AHP-Based Evaluation Method for Vocational Teacher's Competency Standard,." Int. J. Inf. Educ. Technol. **12**(2), 157–164 (2022)
7. İnce, M., Yiğit, T., Işık A.H.: AHP-TOPSIS Method for Learning Object Metadata Evaluation. Int. J. Inf. Educ. Technol. **7**(12), pp. 884–887 (2017)
8. Jiang, H., Liu, Y.: Construction of teaching quality evaluation system of higher vocational project-based curriculum based on CIPP model. Int. J. Inf. Educ. Technol. **11**(6), 262–268 (2021)
9. Xiao, B., Wei, M., Mingshe, D.: Construction and empirical analysis of the evaluation index system for majors set at transportation vocational colleges. Int. J. Inf. Educ. Technol. **9**(11), 778–783 (2019)
10. WANG Lei. Research on-campus security risk assessment based on multi-criteria decision. System Science and Mathematics. 159–170 (2021)
11. Bang, F.E.N.G.: On the preventive assessment of emergency risks in primary and secondary schools. Chin. J. Educ. **11**, 67–72 (2015)
12. Liu, B., Yao, K., Zhao, Z., Ding, S., Chen, H.: Research on the evaluation model of graduate employment prospects. Int. J. Inf. Educ. Technol. **10**(3), 191–195 (2020)
13. LI Penglin, BAO Ting, LI Wei. Research and application of comprehensive evaluation model of campus security. J. Zhejiang Univ. Technol., **49**(04):368–373+383 (2021)
14. Turner, C., Astin, F.: Grounded theory: what makes a grounded theory study? Eur. J. Cardiovasc. Nurs. **20**(3), 285–289 (2021)
15. ZHUANG Yue, LIANG Xiaoxiao. The resilience of urban public health system facing epidemic crisis . China Saf. Sci. J. **2**(02):167–175 (2022)

16. YE Gui, LI Xuezheng, XIANG Qingting, et al. Research on the formation mechanism of construction workers' intentional illegal behavior motivation. China Saf. Sci. J. **30**(08):12–17 (2020)
17. ZHANG Qianqian, ZHANG Yibing, DING Rijia. Macro-security risk warning of regional special equipment based on regulatory perspective. J. Saf. Environ, **16**(02):397–405 (2020)
18. Trivedi, A., Jakhar, S.K., Sinha, D.: Analyzing barriers to inland waterways as a sustainable transportation mode in India: a DEMATEL-ISM based approach. J. Clean. Prod. **295**, 126301 (2021)
19. Priya, S S, Priya, M S, Jain, V, et al. An assessment of government measures in combatting COVID-19 using ISM and DEMATEL modelling. Benchmarking: An International Journal, (2021)
20. Hassan, I.U., Asghar, S.: A framework of software project scope definition elements: an ISM-DEMATEL approach[J]. IEEE Access **9**, 26839–26870 (2021)
21. LI Hu Jun, CHEN Huihua, CHENG Baoquan, et al. Research on the formation model of subway construction safety atmosphere based on fuzzy ISM-DEMATEL. J. Railw. Sci. Eng., **18**(8):2200–2208. (2021)
22. Tie, Y.C., Birks, M., Francis, K.: Playing the game: A grounded theory of the integration of international nurses. Collegian **26**(4), 470–476 (2019)
23. Rieger, K.L.: Discriminating among grounded theory approaches. Nurs. Inq. **26**(1), e12261 (2019)
24. NI Guodong, LI Huaikun, CAO Mingxu, et al. Research on the inducing factors and intervention countermeasures of unsafe behaviors of new generation construction workers. Saf. Environ. Eng. 12–20 (2022)

Generation of Course Prerequisites and Learning Outcomes Using Machine Learning Methods

Polina Shnaider[✉], Anastasiia Chernysheva, Maksim Khlopotov, and Carina Babayants

ITMO University, 49 Kronverkskiy Pr., Lit. A, 197101 St. Petersburg, Russia
`polina.in.tech@gmail.com`, `{avchernysheva,khlopotov}@itmo.ru`,
`karina.babayants@mail.ru`

Abstract. The paper addresses the problem of academic course prerequisites and learning outcomes generation in learning analytics systems. For prerequisites generation, collaborative filtering, i.e., ALS algorithm for Matrix Factorization, is used. For learning outcomes generation, the study discusses an approach based on Computational Linguistics data extraction methods and content-based filtering to recommend potential outcomes. The recommendation mechanisms are designed to be implemented in the Educational Program Maker service for working with education process elements. The study's primary goal is to simplify, formalize and speed up the course development process. Implementation of the approach will make it possible to build unambiguous interdisciplinary connections, identify the closest intersections of the curriculum courses, and build individual learning pathways.

Keywords: Learning analytics · Educational data mining · Computational linguistics · Machine learning

1 Introduction

To date, syllabi are usually used to describe the content of a course in higher education. The document contains a textual descriptive part and may contain the author's keywords indicating what basic concepts are covered in the course and, accordingly, highlighting those things that the student will master from studying the academic course. When preparing a description, recommending the generated core concepts based on the contents already compiled by the author will simplify the process and help bring all the keywords to a unified form. The latter is essential in establishing links between disciplines.

ITMO University uses Educational Program Maker [1] for the syllabi and other educational program-related documents creation. This paper presents a model for prerequisites and learning outcomes generation using machine learning methods.

Educational data mining discovers previously unknown facts about the educational process and its participants to support decision-making. One of the possible decision support tools is developing a recommender system, considering the characteristics of the subject area. The most used recommender systems are collaborative filtering systems, content-based methods, and models based on knowledge. Collaborative filtering systems are used to pre-generate data based on known preferences (assessments) of user groups to predict the unknown preferences of other users. This study describes a collaborative filtering algorithm to generate prerequisites for the syllabi and content-based filtering to generate learning outcomes.

This paper is arranged as follows: Sect. 2 refers to the related work in recommendation systems for learning analytics, Sect. 3 discusses prerequisites generation, and Sect. 4 presents the learning outcomes generation approach. We conclude this paper with an overview of the proposed approach's benefits and limitations along with the future work.

2 Related Work

In recent decades, the evolution of education research has focused on learning analytics, which provides crucial information and can make valuable predictions about the study process. Machine learning methods are widely applied in the learning analytics field. Paper [2] discusses a detailed study on learning analytics by categorizing based on prediction algorithms, data set used, and factors prioritized for prediction. Nowadays, almost every aspect of education or planning of the educational process uses prediction models. The approach in [3] helps discover the relationship between course characteristics and their prerequisites and the resulting students' failure or success.

Liu Q. et al. focus on entity extraction problems and address overlapping entity relation recognition and multiple entity relation extractions across sentences. They propose an entity relations extraction method based on sequence annotation by implementing a probabilistic model for text representation [4].

Ahera S. et al. present a data recommendation system in E-Learning employing Machine Learning and based on historical data. The authors show how data mining techniques such as clustering and association rule algorithms help generate recommendations for newly enrolled students [5].

While most studies such as [3, 6, 7] investigate relations between course prerequisites and student academic success, course prerequisites themselves might be a point of scientific interest. This work is devoted to building recommendation systems for generating relevant curricula prerequisites and learning outcomes using machine learning algorithms.

3 Prerequisites Generation

A recommendation system for generating course prerequisites can utilize various techniques, and one of the most straightforward approaches is finding a similar course and using its entry requirements. It is supposed that study courses with identical learning outcomes might have common prerequisites. Some data preprocessing was made to enable comparisons between curricula: each course's learning outcomes were represented as a binary vector of parameters—$\{x_1, x_2, \ldots, x_n\}$, where x_i takes a value of 1 if the course's outcomes include specific skill and 0 otherwise (Fig. 1).

After data preprocessing, we possess N vectors (courses) having the size of K (all possible study skills in the system). Subsequently, we can compute the dot product (1) between all items and find the most similar courses based on learning outcomes. In (1) a is the first vector, b is the second, n is the dimension of vector space, a_i is the component of vector a, and b_i is the component of vector b.

$$a \cdot b = \sum_{i-1}^{n} a_i \cdot b_i \tag{1}$$

Generally, the prerequisites of the most identical pair of courses possess the most intersection compared to other curricula. For instance, Fig. 2 depicts two different Math courses with their prerequisites. These courses were determined to be the most similar after calculating the dot product.

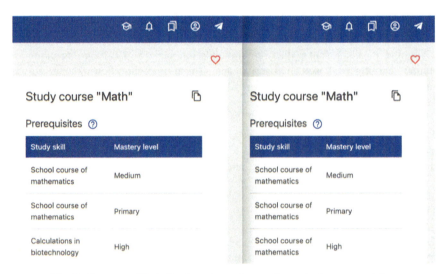

Fig. 2. Courses with similar learning outcomes have similar prerequisites.

However, such a naive approach is not practical since course prerequisites appear at the beginning of filling in information about the study course. Consequently, very little is known about course contents, and learning outcomes cannot be used in the model. It should be noted that the system has approximately 17,000 learning skills used in prerequisites or learning outcomes, which leads to a large size of feature vectors. Consequently, the algorithm calculating dot product between all entities having $O(n^3)$ time complexity does not seem acceptable regarding data size. Therefore, it is aimed to develop an algorithm that will predict relevant prerequisites with no additional info such as learning outcomes, study literature, or syllabus sections and have reasonable computation time.

3.1 Matrix Factorization

Since the dot product calculation is inefficient for computing large feature vectors, an alternative approach was considered—matrix factorization. Matrix factorization is a class of collaborative filtering methods used in recommender systems. Collaborative filtering applies algorithms to filter data from user reviews to make personalized recommendations (Fig. 3) for users with similar preferences [8].

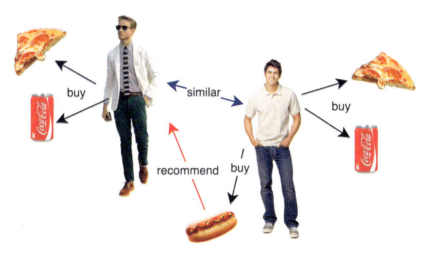

Fig. 3. Demonstration of collaborative filtering idea.

This mechanism is frequently used to recommend digital content or products on the web: streaming services offer movies of potential interest; marketplaces suggest goods customers purchase together. Collaborative filtering is actively applied in educational recommendation systems. In [5] was developed a system that recommends a course to the student based on the choice of other students for a particular set of Moodle courses. For forecasting, the authors of [5] used Simple K-means clustering and the Apriori association rule algorithm.

Matrix factorization is widely applied for making predictions in the learning analytics field. For instance, in [9, 10] factorization approach was used to predict students' marks in study courses.

Regardless of the subject area, all the users' interactions with every item can be represented as a user-item interaction matrix (Fig. 4).

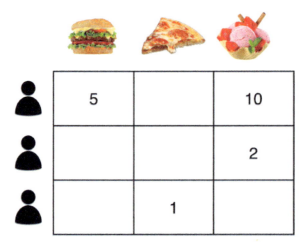

Fig. 4. User-item interaction matrix.

The idea of matrix factorization is to take a vast user-item interaction matrix and decompose it into the product of two lower dimensionality rectangular matrices $R = U \times V$. The user-item interaction matrix is usually sparse; on average, users have interacted with a small number of items, so most of the matrix cells have zeros. In the beginning, matrices U and V are initialized with random values. Then, step by step, the cell-values of the matrices U and V are fitted to obtain, as a result of multiplication, the original matrix R (Fig. 5). During the factoring process, empty cells are filled with non-zero values, and based on their scores, users can be offered corresponding items.

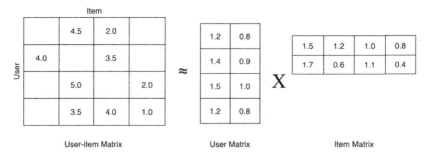

Fig. 5. Matrix factorization principle.

In the discussed task, the user-item matrix consists of course creators (users) and the frequency of study skills they used in their courses (items)—in prerequisites or learning outcomes (Fig. 6).

study skills

1	0	3	0
0	0	0	1
0	1	0	1
1	0	3	0
1	2	0	4

(course editors)

Fig. 6. A user-item matrix to recommend course prerequisites, cells contain the frequency of using each study skill by the editor in all his courses.

We assume that editors develop courses in related subject study areas—such as only physics or only programming. For instance, algorithms courses and high-load programming might require knowledge of Java programming language or experience with system control version Git. If the editor develops course programs in diverse subject areas, such as foreign languages and physics, he will receive recommendations for prerequisites from all subject areas, proportionally to the number of created curricula.

Since the user-item matrix is large and sparse (~1000 users and ~17 000 items, most of the cells contain zeros), for matrix factorization was used ALS (Alternating Least Squares) method. One of the primary features of the ALS algorithm is that it iteratively alternates between optimizing rows and fixing columns and vice versa. Each iteration is done to arrive closer to the original matrix [11] (Fig. 7).

Algorithm 1 ALS for Matrix Completion

1: Initialize U, V
2: **repeat**
3:     **for** $i = 1$ to n **do**
4:         $u_i = (\sum_{r_{ij} \in r_{i\bullet}} v_j v_j^T + \lambda I_k)^{-1} \sum_{r_{ij} \in r_{i\bullet}} r_{ij} v_j$
5:     **end for**
6:     **for** $j = 1$ to m **do**
7:         $v_j = (\sum_{r_{ij} \in r_{\bullet j}} u_i u_i^T + \lambda I_k)^{-1} \sum_{r_{ij} \in r_{\bullet j}} r_{ij} u_i$
8:     **end for**
9: **until** convergence

Fig. 7. Pseudocode for ALS matrix factorization algorithm.

After executing the factorization algorithm and filling missing cells with data, we can make top N prerequisite recommendations with the highest scores for the user. For example, Fig. 8 shows study skills recommendations with their relevance scores for author of two courses: Digital Journalism and Intercultural Communication.

Study skill	Score
Research Article	0.937042
Grant Application	0.910047
Academic presentation	0.896518
Effective Presentation Techniques	0.876185
Storytelling	0.870072
User interface	0.866035
Organization of scientific research	0.861377
Planning	0.855823
Usability	0.854797
English language	0.854328

Fig. 8. Top 10 study skills recommendations with their scores.

4 Outcomes Generation

The current study proposes an algorithm for outcomes generation and recommendation based on the previous approaches discussed in [12, 13]. The leading idea behind the approach is that any scientific or educational entity—textbook, manual, or syllabus—can be described by a small set of keywords that reflect subject area ground concepts. Fundamental ideas can be extracted from a summary of an entity to create equal conditions for each of them: the full text is not always available or exists. Existing approaches to keywords extraction mostly rely on the words' frequency or syntactic layout analysis with network models. It works satisfactorily for extended texts but does not give an acceptable result for brief texts of incomplete sentences. This problem was considered in [12] and applied to scholarly articles summaries in English. However, the present research focuses primarily on the Russian language; thus, algorithms applicable for English are no longer helpful.

As the analysis here concentrates on syllabi contents that usually have a strict structure and encourage short and conceptual sentences, the approach to essential entities extraction is similar to the one proposed in [13]. The general idea was based on the assumption that if the author indicates a particular concept in the course title, section, or topic, it will be studied within the course, and it should be considered a learning outcome. Having more mentions results in a higher position in the output list. Here arise multiple problems. First, concepts can be written in different languages, complicating their semantic comparison, and [13] addressed that problem by implementing USE model [14]. Second, titles can be elongated and contain more than one concept or include none. Third, each course should be described by an equal and small number of key concepts—on average, no less than ten and no more than 20—so that the future model for identifying interdisciplinary connections stands balanced and not overloaded. The situation is such that an author can describe his course extremely briefly or with excessive details. The latter is solved by introducing hints for the author about the maximum number of keywords, but the first case requires additional study of the subject area.

The learning outcomes generation process, considering stated problems, has three steps:

1. Extracting and suggesting concepts explicitly indicated in sections and topics titles. Suitable constructions will be those that satisfy several syntactic rules, no more than three meaningful words in length.
2. Extraction of concepts not explicitly identified in long titles follows the same principles described in paragraph 1.
3. Recommending learning outcomes from courses with similar contents by constructing courses' textual embeddings.

Implementation of the proposed approach offers three blocks of potential learning outcomes: unambivalently distinguished in the headings, extracted by analyzing the syntax, and recommended from similar courses. After that, the author can choose outcomes that suit his course from these three blocks.

4.1 Keyphrases Extraction

Two first blocks organize concepts available in the course description headings. Here, the concept is a 2–3 words keyphrase that is consistent from the perspective of the Russian language. N-grams with higher N values and unigrams are out for the following reasons:

- there are too many stopwords among unigrams.
- unigrams are usually parts of relevant bigrams or trigrams.
- for N-grams, where $N > 3$, the variability of consistent allowable structures rises sharply along with the number of prepositions that complicate the final keyphrase normalization.

Thus, the problem of concepts extraction narrows down to detecting and normalizing bigrams and trigrams that satisfy one of the admissible grammatical structures shown in Fig. 9.

For the English language, syntactic rules are not that inflexible. The main rule is the same as for concepts in Russian—every keyphrase must have a noun as a determinative part of speech. Others are considered not valid by default [12]. Syntactic roles in the original sentence should be regarded if possible. However, existing natural language processing tools do not always successfully solve this problem. ruBERT [15] showed the best results, followed closely by natasha [16]. Natasha was chosen for implementation as its model loading is less time-consuming. In terms of data preprocessing, the following tools were used: razdel [16] and NLTK for tokenization; pymorphy2 [17] and spacy for lemmatization and inflection; Yandex.Translate API [18] for language identification.

For the second block of concepts formation, words compatibility evaluation is necessary. That is performed by normalized pointwise mutual information (NPMI) score calculation (2). Mutual information is a measure of the information overlap between two random variables. Pointwise mutual information indicates how much the actual probability of a two words co-occurrence varies from what we would expect it to be based on

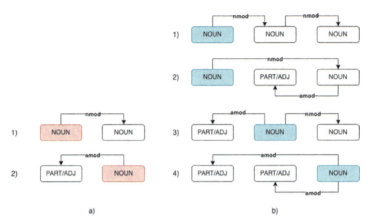

Fig. 9 Syntax structures admissible for keyphrases: a) bigrams, b) trigrams.

the probabilities of these words' occurrence and the assumption of their independence. Normalized PMI metric is less sensitive to the words occurrence frequency.

$$NPMI(x, y) = (\ln \frac{p(x, y)}{p(x)p(y)})/(-\ln p(x, y)) \qquad (2)$$

When two words only occur together, NPMI is 1; when they are distributed as expected under independence, NPMI equals 0 as the numerator is 0; eventually, when two words occur separately but not together, NPMI approaches –1 [19]. The concepts extraction algorithm for the first two blocks is depicted in Fig. 10.

Algorithm 2 Keyphrase Extraction

1: $block_1 \leftarrow \emptyset$
2: $block_2 \leftarrow \emptyset$
3: $S \leftarrow sentencize(T)$ ▷ T is course content
4: **for all** $sentence \in S$ **do**
5: $\quad s \leftarrow preprocess(sentence)$
6: $\quad G \leftarrow \{bigrams(s), trigrams(s)\}$
7: $\quad$ **if** $tokenize(sentence) \in G$ **then**
8: $\quad\quad block_1 \leftarrow block_1 \cup \{sentence\}$
9: $\quad$ **else**
10: $\quad\quad$ **for all** $g \in G$ **do**
11: $\quad\quad\quad$ **if** $check\_syntax(g) = TRUE \cap NPMI(g) \geqslant 0.5$ **then**
12: $\quad\quad\quad\quad block_2 \leftarrow block_2 \cup \{normalize(g)\}$
13: $\quad\quad\quad$ **end if**
14: $\quad\quad$ **end for**
15: $\quad$ **end if**
16: **end for**

Fig. 10. Pseudocode for keyphrases extraction algorithm.

4.2 Similar Course's Keyphrases Recommendation

In the third step, embeddings are built to find the most similar course by cosine similarity. Here, we use Word2Vec word embeddings. In Word2Vec, we have a sizeable unsupervised corpus (i.e., all information we can retrieve from the contents of each course), and for every word in the corpus, we try to predict it by its given context (CBOW) or predict the context given a specific word (Skip-Gram). Word2Vec is a neural network with one hidden layer and an optimization function of Negative-Sampling or Hierarchical Softmax [20]. During the training phase, we iterate through the tokens in the corpus and look at a window of size four and a minimum token occurrence of 1. Then we obtain word embeddings of size 300. As course embeddings maintain the size of word embeddings, such a small size allows quick iterations through the corpus to compute the cosine similarity between each pair of disciplines. Course embedding is the mean of phrase's Word2Vec embeddings weighted with TF-IDF for each word and normalized by the standard deviation (Eq. 3):

$$embedding(X) = (\sum_{i=1}^{N} \frac{W2V(x_i) \cdot TFIDF(x_i)}{\sigma(W2V(x_i))})/N \qquad (3)$$

Thus, we consider course context and words occurrences, while normalization by the standard deviation delivers higher quality recommendations. The concepts recommendation algorithm for the third block is illustrated in Fig. 11.

Algorithm 3 Keyphrase Recommendation

1: $block_3 \leftarrow \varnothing$
2: $M \leftarrow Word2Vec(C)$ ▷ M is a Word2Vec model
3: $X \in M$ ▷ X is a set of word embeddings from M
4: **function** EMBED(a, B) ▷ a is a course content, B is a set of word embeddings
5:     **return** $(\sum_{i=1}^{N} \frac{a_i(B) \times TFIDF(a_i(B))}{\sigma(a_i(B))})/N$
6: **end function**
7: $E \leftarrow \text{EMBED}(C, X)$ ▷ E is a set for embeddings of each course in C
8: **function** POTENTIALKEYPHRASES(x, Y) ▷ x is a random course, Y is a set of course embeddings
9:     $i = argmax(S_C(Y(x), Y))$ ▷ i is an index of the closest embedding to c
10:     **return** $C_i[Keyphrases]$
11: **end function**
12: $K \leftarrow \text{POTENTIALKEYPHRASES}(c, E)$
13: $block_3 \leftarrow block_3 \cup K$

Fig. 11. Pseudocode for keyphrase recommendation algorithm.

The presented way to create course embedding during manual testing (~750 courses) showed better results in predicting only one most similar course than USE embeddings. However, it needs formal evaluation after implementation and data annotation by users. USE model is still utilized for multilingual synonym concepts detection.

Figure 12 shows keyphrases extraction and recommendation results for the "Information Technologies in Technosphere Safety" course divided by blocks and translated into English.

In the example, the contents are too short, so the recommendations may not be exact though they represent the main idea of the course stated by the author, data analysis. The first two blocks consist of the highest possible number of keyphrases; outcomes recommended in the third block are from the "Statistical Methods in Innovation Management" course. These two courses are 85% similar by the contents. The algorithm recommended 16 phrases as potential outcomes. The author can accept all recommendations, select only a few, edit them or add something else.

CONTENT	Application packages for data analysis. Spreadsheets. Packages for engineering and scientific calculations. Geoinformation systems.
BLOCK1	Spreadsheets, Geoinformation systems.
BLOCK2	Analysis program.
BLOCK3	Regression analysis, Paired linear regression, Multiple regression analysis, Descriptive statistics, Frontier Russia, Statistical prediction series, Scientific research, Research and development, Innovation activity, Methodological basis, Science statistics, Intellectual activity, Basic principle of visualization.
TOTAL	16 phrases

Fig. 12. Recommended outcomes for the "Information Technologies in Technosphere Safety" course.

5 Results and Discussion

For prerequisites generation, the ALS algorithm has good convergence and works fast on the vast and sparse matrices (computation time ~30 s using python). Moreover, matrix factorization does not require additional data about the course and can make quality predictions based on users' behavior patterns.

However, the matrix factorization approach based only on study skills usage frequency does not distinguish subject areas. Figure 13 represents sample prerequisites predictions for two users: user A developed courses only in Optics, while user B developed curricula in both Informatics and Innovation subject fields.

Recommendations for user A include both general Physics skills, such as "Subject and tasks of Physics", "Proposing a working hypothesis experiment", "Mathematical apparatus", "Use of integrated research methods" and area-specific skills, such as "Wave optics", "Solid state physics" and "Materials science". These study skills with high probability would be relevant for designing another course in Physics.

Yet, recommendations for user B present an integrated set of study skills and might be less appropriate regarding a particular subject area. Generated skills include numerous specific Legal and Innovation terms such as "Legal basis of intellectual property", "Intellectual property in the Cleantech Center" or "Principles of Corporate Management of Intellectual Property in the Cleantech Sector".

Prediction accuracy can be improved if the user is asked about the subject area at the beginning of filling in the course data, and then the corresponding subject area prerequisites will have more weight.

The learning outcomes generation approach suggests a decent amount of quality keyphrases that can be used by the author while developing an academic course though there are several weaknesses:

- strong dependence on the quality of the available text.
- ambiguity in the part of speech identification (for instance, the word "data" in Russian is more often referred to by morphological analyzers as a participle than a noun).
- unpredictable technical difficulties in processing and inflecting foreign phrases if authors write not in Russian or English.
- inaccuracy in determining the syntactic role of a word in a sentence.

id	name
1022	Wave optics
9997	Solid state physics
14427	Subject and tasks of physics
14429	Division of methods of knowledge
14431	Proposing a working hypothesis
14432	Experiment
14434	Mathematical apparatus
20266	Use of integrated research methods
20750	Materials Science

id	name
427	Informatics
1077	Information systems and technologies
8354	Legal basis of intellectual property
18366	Economics of space
20917	Organization and management of intellectual property results
20918	Statistical Methods in Innovation Management
20920	Intellectual property in the Cleantech sector
20921	Principles of Corporate Management of Intellectual Property in the Cleantech Sector
20922	Development and implementation of patent strategies in the Cleantech sector

Fig. 13 Prerequisites recommendations for user A (left), who developed courses only in optics, and user B (right), who developed courses in the field of computer science and innovation.

6 Conclusion

In the paper, we proposed and described implementation of two separate keyphrase recommendation systems. The first one generates academic course prerequisites using ALS algorithm, while the other extracts relevant learning outcomes for a specific course using computational linguistics methods. In the Results and Discussion section, we indicated the most crucial content-related advantages and drawbacks of the approaches. Both suggested algorithms demonstrated acceptable time consumption and adequate level of recommendations accuracy on the manual evaluation. However, they still need to be tested in situ and formally evaluated.

Therefore, our future work will be focused on the two primary areas. First one will be algorithms' implementation into the Educational Program Maker. The following is important, since testing and implementing the algorithms would allow additional data collection and training models for prerequisites and learning outcomes generation.

The second step would be algorithms' retraining using the gathered feedback from the UI. Additional training could refine user preferences and adjust diverse recommendations in prerequisites generation for users developing courses in various subject areas.

Regarding the generation of prerequisites, it is planned to add more parameters to the algorithm, such as the level of mastery of the educational skill and information about related courses in curricula programs. These aspects are intended to help improve the quality of entity predictions.

Finally, after practical evaluation, we plan to combine two current algorithms into a hybrid recommendation system to increase overall performance and recommendations accuracy among adjusting user experience.

References

1. Educational Program Maker. https://op.itmo.ru/. Last accessed 28 Feb. 2022
2. Sama, R., Thamarai, L., Dr. Paul, P. Victer.: A survey on predictive models of learning analytics. Proc. Comput. Sci. 167, 37–46 (2020)
3. Talbi, O., Chelik, N., Ouared, A., Ali, N.: Additive explanations for student fails detected from course prerequisites. In: International Conference of Women in Data Science, pp.1–7. Taif University (WiDSTaif) (2021)

4. Liu, Q., Jia, X., Yang, W., Tu, F., Wu, L.: Research on entity relation extraction based on BiLSTM-CRF classical probability word problems. In: 13th International Conference on Education Technology and Computers. Association for Computing Machinery, pp. 62–68. New York, NY, USA (2021)
5. Ahera, S.B., Lobo, L.M.R.J.: Combination of machine learning algorithms for recommendation of courses in E-Learning System based on historical data. Knowl.-Based Syst. **51**, 1–14 (2013)
6. McMillan-Capehart, A., Adeyemi-Bello, T.: Prerequisite coursework as a predictor of performance in a graduate management course. J. College Teach. Learn. (TLC) 5(7) (2008)
7. Krol, Ed S et al.: Association between prerequisites and academic success at a Canadian university's pharmacy program. Am. J. Pharm. Educ. 83(1) (2019)
8. Koren, Y., Bell, R., Volinsky, C.: Matrix factorization techniques for recommender systems. Computer **42**, 30–37 (2009)
9. Almutairi, F., Sidiropoulos, N.D., Karypis, G.: Context-aware recommendation-based learning analytics using tensor and coupled matrix factorization. IEEE Journal of Selected Topics in Signal Processing, pp. 1–10 (2017)
10. Jembere, E., Rawatlal, R., Pillay, A.W.: Matrix factorisation for predicting student performance. In: 7th World Engineering Education Forum (WEEF), pp. 513–518 (2017)
11. Hu, Y.F., Koren, Y., Volinsky, C.: Collaborative filtering for implicit feedback datasets. In: Proceedings of the IEEE Int'l Conference Data Mining (ICDM 08), IEEE CS Press, pp. 263–272 (2008)
12. Chernysheva, A., Khlopotov, M., Zubok, D.: Subject area study: keywords in scholarly article abstracts graph analysis. In: CEUR Workshop Proceedings, pp. 155–166 (2021)
13. Koshkareva, M., Khlopotov, M., Chernysheva A.: The development of learning outcomes and prerequisite knowledge recommendation system. Association for Computing Machinery, New York, pp. 1–6 (2021)
14. Yang, Y., Cer, D., Ahmad, A., Guo, M., Law, J., Constant, N., Abrego, G.H., Yuan, S., Tar, C., Sung, Y.-H., Strope, B., Kurzweil, R.: Multilingual universal sentence encoder for semantic retrieval. In: Proceedings of the 58th Annual Meeting of the Association for Computational Linguistics: System Demonstrations, pp. 87–94 (2020)
15. Kuratov, Y., Arkhipov, M.: adaptation of deep bidirectional multilingual transformers for russian language (2019)
16. Natasha. https://github.com/natasha. Last accessed 28 Feb. 2022
17. Morphological Analyzer pymorphy2. https://pymorphy2.readthedocs.io/en/latest/. Last accessed 28 Feb. 2022
18. Yandex.Translate API. https://yandex.ru/dev/translate/. Last accessed 28 Feb. 2022
19. Bouma, G.: Normalized (Pointwise) mutual information in collocation extraction. Proc. Ger. Soc. Comput. Linguist 31–40 (2009)
20. Mikolov, T., Sutskever, I., Chen, K., Corrado, G., Dean, J.: Distributed representations of words and phrases and their compositionality. In: Proceedings of the 26th International Conference on Neural Information Processing Systems, vol. 2(13). Curran Associates Inc., Red Hook, NY, USA, pp. 3111–3119 (2013)

Learning Factors for TIMSS Math Performance Evidenced Through Machine Learning in the UAE

Ali Nadaf, Samantha Monroe[✉], Sarath Chandran, and Xin Miao

Alef Education, Abu Dhabi, UAE
samantha.monroe@alefeducation.com

Abstract. Understanding how the UAE K12 education system performs with data-driven evidence is key to inform better policy making to support UAE vision to upskill human capital growth for its economic transformation. In this study, we investigate the potential of using machine learning techniques to understand key learning factors contributing to UAE student math performance on the TIMSS 2019 assessment. Due to the fact that learning factors co-exist and interact with one another, we explore the SHapley Additive exPlanations (SHAP) approach to explain the complexity of the model. The results highlight the importance and contributions of each learning factor and uncover the relationships between the learning factors. Understanding key learning factors and identifying evidence-based intervention opportunities will help policymakers with informed education intervention designs to improve student mathematics learning, in order to improve UAE student TIMSS math performance over the long run.

Keywords: Learning factor analysis · TIMSS 2019 · UAE K12 Education · SHAP values · Machine learning

1 Introduction

Education is one of the basic elements of human capital theory, which predicts that increased knowledge and skills through education will yield improved economic returns for both individuals and societies [1]. Studies have indicated that there is a strong correlation between the country's economic growth and learner educational achievement [2–6]. A country's investment in education is therefore a critical first step to transfer values to future generations [3, 5].

Many countries strive hard to assess their quality of education to improve student academic achievement [3, 7], while comparing their student achievement on a national level against that in other participating countries in international benchmark tests like TIMSS (Trends in International Mathematics and Science Study) and PIRLS (Progress in International Reading Literacy Study) [7–10]. These studies assess students' academic achievement at the international level and are widely accepted as baseline measurements of student academic achievement in each relevant subject, while also providing a system-level diagnosis of a country's educational attainment in different disciplines and highlighting important learning factors to guide future improvement [9]. Thus, understanding how various learning factors contribute to students' academic achievement plays a pivotal role for decision-makers in the field of education. However, students' performance is an outcome of compounding effects from various stakeholders from the education system (e.g. teachers, students, parents, etc.), so categorising close learning factors associated with academic achievement is a complex task due to the range of learning factors and the inter-relationship that exists between them [11–13]. This elevates the necessity of a machine learning model, which is capable of fine-tuning the very nature of the learning factors contributing to students' academic achievement. Thus, the most recent information available on learning factors of student academic achievement can guide policymakers to make evidence-based policy decisions on various contributing aspects that are important to student learning.

TIMSS was developed by the International Association for the Evaluation of Educational Achievement (IEA) to measure student achievement in Mathematics and Science at Grades 4 and 8 (Years 5 and 9). It is intended to measure how well students have learned mathematics and science curricula in participating countries and students from the United Arab Emirates have participated since 2011 (fourth assessment cycle). In the most recent cycle (seventh assessment cycle), over 48,000 students from about 625 public and private schools across the United Arab Emirates participated in the TIMSS 2019 study. The ranking for UAE grade 4 students in mathematics is at 43rd with 481 points and in science at 31st with 473 points, while in eighth grade it ranked 26th in both mathematics and science with 473 points each. The improvement since the first assessment has been observed for fourth-grade students with 29 points in mathematics and 22 points in science. However, eighth-grade students had a minor improvement of 8 points in mathematics and a decrease of 4 points in science. TIMSS has supported national curriculum and K12 education policy in the UAE to make strategic reforms in accordance with the National Agenda in 2010–2022 to raise the students' performance, address weakness in certain domains [14].

TIMSS was "designed to provide countries with information about their students' mathematics and science achievement that can be used to inform evidence-based decisions for improving educational policy and practice" [15]. The comprehensive TIMSS datasets enable researchers to link students' contexts for learning to their math academic achievement, resulting in TIMSS data being a rich and valuable source to measure an education system's progress. Learning factors are an inherent part of the TIMSS survey design—the survey design includes understanding holistically the performance of participating students and countries while including all stakeholders in student learning through surveys completed by students, parents or guardians, teachers, principals, and national curriculum specialists. Understanding influential learning factors and identifying evidence-based intervention opportunities help policymakers to recommend better education strategies to prepare students for the upcoming TIMSS assessments, the next of which is in 2023.

Based on the above information, this paper sets out to answer the following research questions using 2019 TIMSS data[1] on student mathematics achievement:

---

[1] Retrieved from the TIMSS 2019 International Database Downloads, https://timss2019.org/international-database/.

(1) What are the key learning factors that contribute to grade 4 and grade 8 student TIMSS mathematics achievement in the UAE?
(2) Is there any difference in the top influential learning factors between 4th and 8th grade for UAE students' mathematics learning? If there is, what is the difference in these factors?

The rest of this paper is organised as follows: in Sect. 2, we review the related literature, in Sect. 3, we discuss the methodology in further detail, Sect. 4 presents the results and discussion, and in Sect. 5, we provide the conclusion of this research paper.

2 Literature Review

Since its formation in 1971, the United Arab Emirates (UAE) has recognized the value of investing in education to drive domestic human capital growth to support the vision of its economic transformation from oil-based to knowledge-based. This vision has been translated into ambitious goals in the education sector, such as UAE students ranking among the top 15 countries in the TIMSS assessment by 2021 [14].

While the UAE's education system has undergone rapid improvement in recent years, there is still work to be done before results and performance are fully on par with students in the same age group in other countries. Thus, the aim of this research is to utilise machine learning techniques and the SHapley Additive exPlanations (SHAP) approach, to gain an understanding of learning factors that influence students' math learning. The findings can be used to ensure all students are provided with the proper learning interventions. We have therefore conducted a literature review which includes broadly two areas: identifying the learning factors contributing towards students' academic achievement and its interpretation and employing potential machine learning techniques, including SHAP, to analyse TIMSS data.

Learning factors contributing to students' academic achievement and its interpretation. There has been an increasing number of studies carried out in identifying the learning factors (features) that contribute to students' academic achievement. These studies have utilised different data mining techniques or models to explore the features based on the interest and purpose of the study and based on the learnings from the past studies, features have been grouped broadly into individual and family-related, school environment, and social environment-related variables [11, 13, 16–19].

Individual factors refer to personal factors to the student, while family-related refers to the student's family details to the extent factors affect the students' learning. Individual characteristics and family-related factors includes age, gender, motivation, self-esteem, and self-confidence in the subject, prior academic performance, socio-economic status of student families, parental involvement, long-term educational aspirations, educational opportunities at home, and student orientation towards learning math/science [7, 8, 10, 11, 13, 17, 18, 20–24]. Home educational resources and student confidence in mathematics are found to be highly important learning factors in students' math academic achievement [9], while location of the school, parent education, gender, school type, and education opportunities were found to be prominent learning factors influencing the TIMSS eighth-grade math performance [5]. Kiray et al. 2015 found that self-concept, confidence, motivation, self-efficacy, anxiety, and attitude are the most influential learning factors of students' academic achievement in science and math subjects and cognitive skills emerged as a strong predictor of students' achievement in international achievement tests across OECD countries [2, 18].

School environment factors are another group of variables influencing student academic achievement. Some examples include curriculum quality, opportunity to learn, attendance, active learning, instruction time, class climate, adaptive instruction (opportunity for collaboration/extent of solving problems on their own), effective learning time, teacher motivation, teacher workload, student-teacher interaction, teacher qualification, teacher pre-service and in-service courses, teacher responsiveness, size of the class, the location of the school, and more [4, 7–11, 13, 17, 18, 20–26]. However, some studies have confirmed the existence of an effect of school-related variables on students' performance [26], while others have stated that there is a lesser contribution or a non-existent contribution [27].

Finally, social factors are linked to other group factors, with some of the social factors reported in the past studies being the location of students, the community of students, cultural heritage, and idiosyncrasies [26, 27].

Statistical and machine learning techniques used by studies. Several studies have been carried out on educational data sets, mainly to uncover useful information pertaining to students' academic achievement. Cardoso et al. 2011 used confirmatory factor analysis and structural equations to study the influence of learning performance on academic achievement, whose results revealed that self-esteem, teacher-student interaction, and student-student interaction have a direct positive influence on learning performance, which in turn has a direct positive influence on academic achievement [11, 28, 29]. Learning performance was defined in this study as students' self-evaluation of acquired knowledge, understanding and skills developed, and their desire to learn more, while academic achievement was described by the grades obtained in the subject [11].

DeFreitas and Bernard 2015 explored partitioning, hierarchical, and density-based clustering techniques such as K-means, BIRCH, and DBSCAN respectively on educational datasets for classification [12]. This is being used subsequently for predictive modelling, evaluation to create groups for better understanding of stakeholders within the educational environment, and exploration, which refers to a clear understanding of the dataset for further analysis.

Filiz and Ersoy 2020 explored algorithms such as k-NN, Naive Bayes algorithm, support vector machines, artificial neural networks, decision trees, and logistic regression to determine the factors influencing students' academic achievement [9]. Lau, Sun, and Yang 2019 utilised Artificial Neural Network (ANN) modelling for the prediction of students' academic achievement [22].

Studies that analysed learning factors for students performance and academic achievement generally employed multiple regression [2, 5, 20], quantile regression [3], logistic regression [25], and decision tree algorithms (J48, is the C4.5 algorithm which is an extension of ID3 algorithm) [13, 16, 18, 21]. Similar studies on TIMSS have used conventional statistical techniques which have a limited ability to explain the contribution of the learning factors on students' performance. Research was found utilising machine learning to discover key learning factors via the TIMSS assessment data, but none utilised the SHAP method or worked with the 2019 dataset [9, 30, 31]. Several studies utilised the SHAP approach to uncover learning factors within the 2018 PISA dataset [32, 33].

3 Methodology

3.1 SHapley Additive exPlanation (SHAP)

In today's research world, machine learning is utilised to solve and unravel complicated challenges. Machine learning is employed in cross-disciplinary research to uncover patterns in data and the SHAP approach is used to comprehend how features of the model are related to its outputs.

SHAP is a coalitional game-theory and a local explanation technique to quantify the contribution of each feature, either positively or negatively, to the model outputs. The SHAP value of a feature is measured by comparing what a model predicts with and without that feature against a baseline value. The technique offers a high level of local interpretability for a model so that each data point has its own SHAP value. The aggregation of these local explanations for each prediction enables us to understand the global structure of the machine learning model [34–36].

In the case of predicting student math academic performance, in the absence of information about a student, it would be most accurate to estimate the mean of the overall scores to predict student math score (baseline value). If one has information about the student's language background, the prediction would change and be more accurate. The prediction would change again if one included data about the students' behaviour. Therefore, there will be a difference in predictions at any stage if a certain feature is omitted or included. Such differences indicate how the specific feature affects the final prediction. In order to measure SHAP values of feature A, one needs to compare the model outputs with the feature A for all the combinations of features other than A.

The locality of explanation of SHAP values represents the contribution of each data point on the model outcomes. For instance, if the SHAP value of the "Home Educational Resources" feature for a student is +10, the value indicates that this feature contributes positively to student math performance and increases it by +10 points. A SHAP value of zero represents no contribution to the model output. The SHAP approach focuses on the contribution of features to model output without attempting to infer causality.

3.2 Model Input—Learning Factors

The TIMSS 2019 International database comprises 4th and 8th-grade students' achievement data in mathematics and science, as well as survey responses regarding factors related to students, teachers, parents, schools, and curricula. Each student participating in the TIMSS 2019 assessment completed a survey, which asked about students' home life, school, basic demographic questions, the home environment, school climate, and students' attitudes about mathematics and science. Parent or guardian questionnaires asked about home resources for supporting student learning, the highest level of education and employment situations within the household, and opinions about their child's school, as some examples. Teacher surveys discussed teacher education, professional development, and experience in teaching, amongst other categories. School questionnaires were completed by principals regarding numeracy and literacy, the socioeconomic status of the school, and the school's culture towards academic achievement. Lastly, the TIMSS 2019 National Research Coordinator within each country was responsible for the Curriculum Questionnaires, which focused on national curriculum policies and practices and the content of the mathematics and science curricula [37].

TIMSS uses a two-stage random sampling process to collect data. The first stage is selecting schools proportionally, and the second stage is selecting classes randomly within sampled schools [38]. This study is mostly based on self-reported questionnaires for UAE participating students.

Features with more than 50% missing values were dropped from the final analysis. Other missing values were imputed using different strategies like mean imputation for numeric features and most frequent (mode) imputation for categorical features with less than 10% missing values. For other categorical features, missing values were replaced with zero. The imputed data was normalised with a standard scalar before it is passed for model training.

Furthermore, features with collinearity values greater than 60% were excluded from the analysis because of their high correlations or overlapping information with other features. In the appendix, we present the complete list of 61 and 46 final input features used in the analysis for grades 4 and 8, respectively, along with their definitions.

3.3 Model Output—Student TIMSS Performance in Mathematics

TIMSS does not estimate a single test score for the math cognitive aspect of each student, but instead estimates a probability distribution for his or her achievement and deduces plausible values (PV) for a student's math performance from this distribution. The results for the student performance are computed by taking a weighted average for each PV. The final math academic achievement score per student is used as the model output [38].

3.4 Boosted Regression Tree Model

The input features are encoded into matrices of 22,163 and 27,342 rows representing the number of 4th and 8th grade UAE students who participated in TIMSS 2019, respectively. We split the data into training and validation sets with a ratio of 80/20.

We have trained a boosted-regression-tree model using the CatBoost package in Python [39]. The CatBoost model was trained with 5000 estimators (regression trees) with a tree depth of 10 and a learning rate of $\eta = 0.05$. Root mean square error (RMSE) was selected as the loss function for the model.

We highlight that linear models, which may appear to offer more interpretable findings at first glance, may not present reality. When nonlinearity in the data becomes significant, a linear model can be less interpretable than a boosted regression tree [36].

4 Results and Discussion

4.1 Model Performance

The results of the training for grade 4 TIMSS data show that the root mean square error (RMSE) is 50.23, the mean absolute error (MAE) is 40.42, and the coefficient of determination (R^2) is 0.71 in the validation set. The RMSE, MAE, and R^2 for grade 8 validation data are 47.09, 37.44, and 0.76, respectively. As a rough rule of thumb suggested by Hair et al., R^2 values of 0.71 and 0.76 can be considered as substantial levels of predictive accuracy [40]. The model performance for both grades 4 and 8 is shown in Fig. 1.

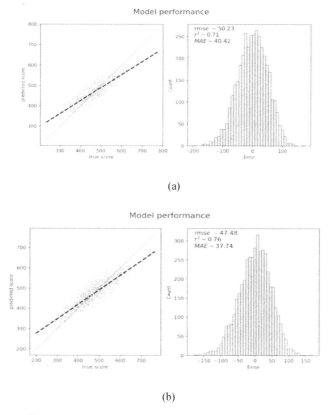

Fig. 1. Model performance metrics for grades (a) 4 and (b) 8

4.2 SHAP Summary Plot of Learning Factor by Importance

We used the SHAP approach to explain the structure of the model built by CATBoost and to elucidate the relative importance of learning factors to student TIMSS math performance. Figure 2 shows the summary plot of the top 19 features in terms of their importance for UAE students. Each dot on the plot represents the SHAP value of a predictive factor for a student.

Among the top 19 important features, four categories are distinguished relevant to mathematics: (1) student related; (2) teacher related; (3) family related; (4) demographics related. Student factors include students' confidence in mathematics, student learning speed, peer comparisons, and language difficulties. Teacher factors include teachers' level of education, their understanding of the curriculum, and student grouping choices. Parent factors can be seen from parental involvement in school activities, parents' opinions on mathematics, parents introducing early numeracy and literacy, and parental involvement in student learning. Lastly, school locality (i.e. urban area, rural, etc.) and regional population size are top features related to demographic factors.

Among these predictive factors, student confidence in mathematics, language difficulties, and measurement and geometry familiarity rank as the top influential factors that contribute to 4th grade student math scores on TIMSS. Meanwhile, instruction time, language difficulties, and independent work are the most influential factors for 8th grade students. This illustrates the consistency of some areas for students, such as how language learning can heavily contribute to academic achievement in mathematics. It must be noted that language difficulties in this context means student difficulty with the language of the TIMSS exam, which can be given in either English or Arabic for UAE students. Measurement and geometry familiarity refers to the percent of students taught measurement and geometry topics either that academic year or prior. Instruction time is how many days per year a school is open for instruction, while independent work is categorised as how often students work on problems on their own during mathematics lessons.

4.3 Discussion

According to Fig. 2(a) for fourth grade, student confidence in mathematics and whether they are taught measurement and geometry topics experience both high SHAP values and high confidence values, delineating a greater probability to get higher scores on the mathematics section of the TIMSS assessment for these students. For Fig. 2(b) for eighth grade students, the amount of time spent on math instruction and teacher education levels similarly result in high SHAP and confidence values, leading to a greater probability of obtaining better mathematics scores on TIMSS. Negative SHAP values for language difficulties for both 4th and 8th grade students indicate a negative contribution of language to TIMSS math scores. Furthermore, we can also see that language difficulties SHAP value distribution is very wide for both grade 4 and grade 8 students (see Fig. 2(a) and 2(b)), which means that the contribution of language to math learning can be extremely positive or extremely negative. In other words, when a student has extremely low language skills, the potential math performance of this student is negatively affected to a great extent, while a positive contribution also exists, but to a lesser extent.

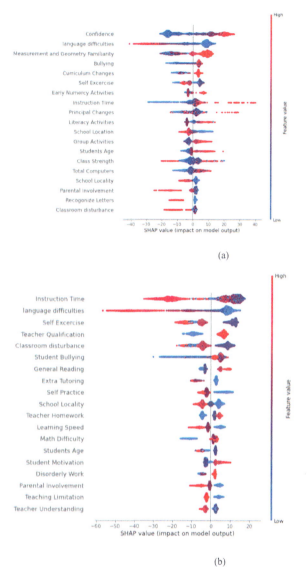

Fig. 2. SHAP features importance for determining student TIMSS performance in Mathematics for grades (a) 4 and (b) 8

Among both 4th and 8th grade students, language difficulties are one of the top factors contributing to students' mathematics scores on the TIMSS assessment, with their display in the figures nearly identical across grades, meaning the impact on students is apparently steady over time. Between both grades of students, language difficulties have a very strong negative contribution towards students' math scores, with a majority of students seemingly facing this issue. However, there is a cluster of students (a bubble in Fig. 2) whose language difficulties have a positive contribution, although to a lesser extent, especially when compared to its potential negative impact on their peers. Language difficulties imply that students have an overall lower ability in language learning—which can include speaking, reading, or writing—for the mathematics instructional language (either English or Arabic), which in turn hinders their mathematics learning. Additionally, based on the widely spread out distribution of SHAP values for language difficulties for both grade 4 and grade 8 students, it is clear evidence of strong heterogeneity when it comes to contribution of language difficulty for student math performance, which requires varied intervention approaches for different groups of students. SHAP dependence plots are used to interpret the contribution of language difficulties on student math performance (see Fig. 3). Each dot in the figure represents a single prediction of student performance. Figure 3 displays a nonlinear relationship between language difficulties and their contribution to student math performance. A similar pattern for the contribution of language difficulty is observed for both grade 4 and grade 8 students. This nonlinear relationship reveals the complexity of the education system and illustrates the importance of using nonlinear machine learning techniques to discover complex relationships between factors that affect learning.

When looking at the most important factors between the two grades, confidence in mathematics is the number one for 4th grade students. For students in which their confidence in mathematics contributes positively to their mathematics score on the TIMSS assessment, it has a moderate to high correlation reflecting in student scores. On the other hand, for students in which their confidence contributes negatively to their TIMSS math score, it has a lesser effect (see Fig. 2). It is also important to note that there is a cluster of students whose confidence in mathematics negatively contributes to their math score, particularly at the lower end of the spectrum, although there is a large range overall, meaning confidence in mathematics has an incredibly varied affect on math scores between 4th grade students.

As observed in Fig. 2(a), for parental involvement in school activities, letter recognition, and classroom disturbance, while a large portion of students experience a slight positive contribution to their math scores from these factors, there are students who experience a very large negative contribution from these three areas. For the TIMSS 2019 assessment, parental involvement in school activities dictates how often parents engage with activities at a school in general, letter recognition refers to a student's ability to recognize letter before attending primary school (i.e. early literacy capabilities), and classroom disturbance is the degree to which student behaviour is an issue for fourth grade students within a given school. These are important features, especially within the UAE context, as the first two highlight parental involvement both during and before students begin primary school. For classroom disturbance, this has been a trend within the UAE school system and has been documented by the researchers during several observations, as well as noted by teachers and principals in discussions. For all three learning factors, their overall contribution to students' math academic achievement is largely negative, signifying their impact on 4th grade students in the classroom.

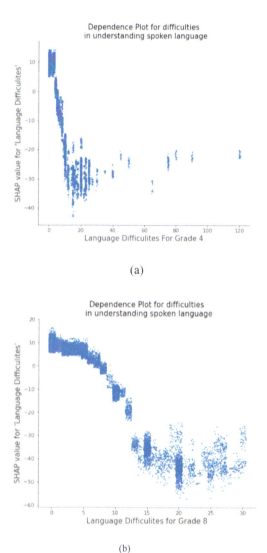

Fig. 3. Dependence plots for language difficulties for (a) grade 4 and (b) grade 8

5 Conclusion

5.1 Summary of Key Findings

Overview of the study. In this research, we evaluated the contribution of learning factors on predicting the TIMSS score of UAE students in mathematics by measuring the factors' contribution with SHAP values.

To uncover key learning factors, we explored and selected learning factors using domain knowledge. Then, we transformed the most relevant variables from TIMSS data and developed a machine learning predictive model. After investigating multiple regression models, we finally found that a boosted regression tree model can generate the optimal model with RMSE of 50.23 and 47.09 for grade 4 and 8 datasets, respectively. This model also returns the highest coefficient of determination (R^2) 0.71 and 0.76 for grades 4 and 8, respectively. These results clearly indicate that we can predict student math performance using learning factors in the TIMSS dataset.

We used SHAP technique to measure a local explanation for the gradient boosted tree algorithms. We aggregated the local explanation to understand the underlying structure of the like computing the feature importance using the mean of absolute SHAP values for each feature. After computing the local SHAP values, the dependence between each pair of features was analysed. The relationship between factors revealed how the factors change students' math cognitive performance.

Findings of the study. Among the top 19 important learning factors for grade 4, confidence in mathematics is identified as the best predictor to be associated with student math ability in the TIMSS assessment. Confidence, also described as self-efficacy, is a student's self-belief in their ability to overcome difficulties to solve mathematical problems This finding is aligned with what Miao et al. found using math diagnostic test results and self-assessment surveys for Abu Dhabi public school students [41].

The other influential factors influencing grade 4 students' math learning are language difficulties and school-related factors such as bullying, learning styles, and independent work. Our findings show that early education activities in literacy and numeracy also contribute positively to student math learning, which is aligned with PISA 2018 results [42, 43].

Top influential learning factors for grade 8 students are almost consistent to the ones for grade 4 students. Language difficulties, learning styles, and independent work are among the top learning factors for math learning for UAE grade 8 students. This concludes that the top learning factors for elementary and middle school learners in the UAE are mostly consistent, which assists policymakers and curriculum developers to design and distribute a connected intervention program to improve student proficiency in mathematics.

Policy implication of the study. One of the implications of the study is to use research to uncover evidence-based findings to inform policymakers' about key learning factors using systematic machine learning and research methodology. Although the factors used for this study were limited to learning variables collected from the TIMSS assessment, the findings point to a number of promising avenues for policy development, future research, and research and development initiatives in the UAE's education sector.

According to the results, for grade 4 students language difficulties and confidence in mathematics play key roles in students' math academic achievement. This makes sense, as student approaches to learning involve the core interplay of the student and the teacher in the presence of the learning content, which is the "Instructional Core". "Increases in student learning occur only as a consequence of improvements in the level of content, teachers' knowledge and skill, and student engagement. If you change any single element of the instructional core, you have to change the other two" [44]. Within the Instructional Core, the student plays a central role. The student directs the information processing inward to have domain-specific judgments about self and task (i.e.

self-efficacy), in this case, confidence in solving math problems, then he or she would develop affective reactions, motivation, and then outward at the required learning activities using metacognitive and cognitive strategies and instructional language to solve mathematical problems [45]. In other words, mathematical content should be readable and understandable for students with different English proficiency levels, requiring learning materials to be equipped with the right level of instructional language. In addition, the role of policymakers in fostering students' language abilities is extremely important by designing proper language programs for students in elementary and middle schools. In such programs, learners' language needs must be identified and effective intervention programs should be designed to help students meet grade level language standards. Moreover, educators' language proficiency to deliver math instructions successfully should be improved through professional development. Educators also need to be mindful of cultivating student confidence to approach math learning in general.

In addition to policy refinement, the study shows the promises of AI techniques to uncover latent learning factors. This is an illustration of how AI algorithms can be used in extracting complex nonlinear relations between learning factors and students' math learning, and explaining the relationships by providing interpretable representation. With the wide use of education technology in education systems, more and more data will be collected. The combination of big data and AI algorithms could potentially provide data-driven insights that traditional education research and reform approaches could not. Additionally, policymakers could be better informed of issues with more precision, hence targeted interventions could be designed.

5.2 Limitations

The main limitation of this study is that the TIMSS 2019 dataset does not cover all possible key learning factors, such as economic and metacognitive factors. Missing that information may impact the outcome of the analysis. However, the TIMSS data is the largest international dataset for math learning for 4th and 8th grade students. Additionally, our findings and results are highly dependent on the TIMSS data quality.

Due to the size of the research and time limitation, we have not explored the detailed impact of each learning factor on students' math learning. Furthermore, the tree SHAP model ensures local explainability of the AI model but it is not model-agnostic. This indicates that the explainability of the model can be changed by a small deviation in the model hyper-parameters.

5.3 Future Work

For future work, one study can be estimating the validity of the model explanation with the support of domain experts, especially when the model is highly non-linear and interpretation of the model is complex. Such models require domain knowledge to interpret results in detail.

An additional study can be conducted to identify cut-off values for each learning factor separating negative SHAP values from positive ones. These cut-off values can be used as a threshold for intervention design and also as a benchmark for decision-makers to monitor and evaluate the progress of students' learning. It is worth exploring the changes in learning factors across different TIMSS assessments and monitoring how the key learning factors have evolved over time for 4th and 8th grade students.

Appendix

The full list of learning factors used to predict grades 4 and 8 students in this analysis are displayed in Tables 1 and 2, respectively.

Table 1. List of features used for grade 4 prediction

Code	Description	Code	Description
ACBGDAS	School discipline-principal	ASDGSCS	Students confident in science
ACBGEAS	School emphasis on academic success-principal	ASDGHRL	Home resources for learning
ACBGLNS	Students enter with literacy and numeracy skills	ASDGICM	Instructional clarity in mathematics lessons
ACBGMRS	Instruction affected by math resource shortage	ASDGSLS	Students like learning science
ACBGSRS	Instruction affected by science resource shortage	ASDGSSB	Students sense of school belonging
ACDGTIHY	Total instructional hours per year	ASDHAPS	Student attended preschool
ACDGSBC	School composition by socioeconomic background	ASDHEDUP	Parents highest education level
ACBG05A	Region population size	ASDHELA	Early literacy activities before school
ACBG05B	School locality	ASDHELN	Early literacy and numeracy activities before school
ACBG06B	Instruction time, minutes	ASDHELT	Early literacy tasks beginning school
ACBG07	Total computers	ASDHENA	Early numeracy activities before school
ACBG14E	Parental involvement, school activities	ASDHENT	Early numeracy tasks beginning school
ACBG15C	Classroom disturbance	ASDHLNT	Early literacy and numeracy tasks beginning school
ACBG18	Principal changes	ASDHOCCP	Parents highest occupation level

(*continued*)

Table 1. (*continued*)

Code	Description	Code	Description
ASBGDML	Disorderly behaviour during math lessons	ASDHPSP	Parents perceptions of their child school
ASBGHRL	Home resources for learning	ATBG01	Years been teaching for teachers
ASBGICM	Instructional clarity in mathematics lessons	ATBG02	Sex of teacher
ASBGSCM	Confidence in mathematics	ATBG03	Age of teacher
ASBGSB	Student bullying	ATBG04	Level of formal education completed for teacher
ASBGSLM	Students like learning mathematics	ATBG10A	Class size
ASBGSLS	Students like learning science	ATBG10B	Number of students in <4th grade>
ASBGSSB	Students sense of school belonging	ATBGEAS	School emphasis on academic success-teacher
ASBHELA	Early literacy activities before school	ATBGLSN	Teaching limited by student not ready
ASBHELN	Early literacy and numeracy activities before school	ATBGSOS	Safe and orderly schools-teacher
ASBHELT	Early literacy tasks beginning school	ATBG09G	Curriculum changes
ASBHENA	Early numeracy activities before school	ATBG11	Language difficulties
ASBHENT	Early numeracy activities before school	ATBM02G	Mixed ability groupings
ASBHLNT	Early literacy and numeracy tasks beginning school	ATDGTJS	Teachers job satisfaction
ASBHPSP	Parents perceptions of their child school	ATDMMEM	Teachers majored in education and mathematics
ASBH06A	Letter recognition	ATDMNUM	Pct students taught number topics
ASBM01	Independent work	ATDSMES	Teachers majored in education and science
ASDAGE	Students age	ATDGTJS	Teachers job satisfaction
ASDG05S	Number of home study supports	ATDMMEM	Teachers majored in education and mathematics
ASDGDML	Disorderly behaviour during math lessons	ATDMNUM	Pct students taught number topics

(*continued*)

Table 1. (*continued*)

Code	Description	Code	Description
ASDGICS	Instructional clarity in science lessons	ATDSMES	Teachers majored in education and science
ASDGSB	Student bullying	ITSEX	Sex of student
ASDGSCM	Students confident in mathematics		

Table 2. List of features used for grade 8 prediction

Code	Description	Code	Description
BCBGDAS	School discipline-principal	BSDG05S	Number of home study supports
BCBGEAS	School emphasis on academic success-principal	BSDGDML	Disorderly behaviour during math lessons
BCBGMRS	Instruction affected by math resource shortage	BSDGICM	Instructional clarity in mathematics lessons
BCBGSRS	Instruction affected by science resource shortage	BSDGICS	Instructional clarity in science lessons
BCDGTIHY	Total instructional hours per year	BSDGSB	Student bullying
BCDGSBC	School composition by socioeconomic background	BSDGSCS	Students confident in science
BCBG05B	School locality	BSDGSLM	Students like learning mathematics
BCBG14A	Teacher understanding of curriculum	BSDGSLS	Students like learning science
BCBG20	Teacher education	BSDGSSB	Students sense of school belonging
BSBGDML	Disorderly behaviour during math lessons	BSDGEDUP	Parents' highest education level
BSBGHER	Home educational resources	BTBG01	Years been teaching for teachers

(*continued*)

Table 2. (*continued*)

Code	Description	Code	Description
BSBGICM	Instructional clarity in mathematics lessons	BTBG02	Sex of teacher
BSBGSCM	Students confident in mathematics	BTBG03	Age of teacher
BSBGSB	Student bullying	BTBG04	Level of formal education completed for teacher
BSBGSEC	Self-efficacy for computer use	BTBG06G	Parental Involvement, Learning
BSBGSLM	Students like learning mathematics	BTBG10	Number of students in the class
BSBGSLS	Students like learning science	BTBG11	Language difficulties
BSBGSSB	Students sense of school belonging	BTBG13E	Disruptive students
BSBGSVM	Students value mathematics	BTBM14	Instruction time, minutes
BSBGSVS	Students value science	BTBGEAS	School emphasis on academic success-teacher
BSBG04	Books at home	BTBGLSN	Teaching limited by student not ready
BSBM15	Independent work	BTBGSOS	Safe and orderly schools-teacher
BSBM18C	Disorderly lessons	BTBGTJS	Teachers job satisfaction
BSBM19A	Mathematics performance	BTDGEAS	School emphasis on academic success-teacher
BSBM19B	Peer comparison	BTDGLSN	Teaching limited by student not ready
BSBM19D	Learning speed	BTDGSOS	Safe and orderly schools-teacher
BSBM19F	Solving difficult problems	BTDGTJS	Teachers job satisfaction
BSBM20A	Mathematics in daily life	BTDMNUM	Pct students taught number topics
BSBM20H	Parents opinion on mathematics	BTDMALG	Percent of students taught algebra topics
BSBM26AA	Homework frequency	BTDMDAT	Percent of students taught data/probability topics
BSBM43BA	Extra tutoring	BTDMGEO	Percent of students taught geometry topics
BSDAGE	Students age		

References

1. Hanushek, E.A., Woessmann, L.: Education, knowledge capital, and economic growth. Econ. Educ. 171–182 (2020). https://doi.org/10.1016/b978-0-12-815391-8.00014-8
2. Hanushek, E.A., Woessmann, L.: How much do educational outcomes matter in OECD countries? Econ. Policy **26**(67), 427–491 (2011)
3. Ibourk, A.: Determinants of educational achievement in Morocco: a micro-econometric analysis applied to the TIMSS study. Int. Educ. Stud. **6**(12), 25–36 (2013)
4. Sandoval-Hernández, A., Białowolski, P.: Factors and conditions promoting academic resilience: a TIMSS-based analysis of five Asian education systems. Asia Pac. Educ. Rev. **17**(3), 511–520 (2016)
5. Sulku, S.N., Abdioglu, Z.: Public and private school distinction, regional development differences, and other factors influencing the success of primary school students in Turkey. Educ. Sci.: Theory Pract. **15**(2), 419–31 (2015)
6. Suri, T., Boozer, M.A., Ranis, G., Stewart, F.: Paths to success: the relationship between human development and economic growth. World Dev. **39**(4), 506–522 (2011)
7. Drent, M., Meelissen, M.R.M., van der Kleij, F.M.: The contribution of TIMSS to the link between school and classroom factors and student achievement. J. Curric. Stud. **45**(2), 198–224 (2013)
8. Bofah, E.A., Hannula, M.S.: TIMSS data in an African comparative perspective: investigating the factors influencing achievement in mathematics and their psychometric properties. Large-Scale Assess. Educ. **3**(1), 1–36 (2015)
9. Filiz, E., Enes, Öz.: Educational data mining methods for TIMSS 2015 mathematics success: Turkey case. Sigma J. Eng. Nat. Sci. **38**(2), 963–77 (2020)
10. Kwak, Y.: An analysis of the Korean science education environment for 20 years of TIMSS. J. Korean Earth Sci. Soc. **39**(4), 378–387 (2018)
11. Cardoso, A.P., Ferreira, M., Abrantes, J.L., Seabra, C., Costa, C.: Personal and pedagogical interaction factors as determinants of academic achievement. Procedia-Soc. Behav. Sci. **29**, 1596–1605 (2011)
12. DeFreitas, K., Bernard, M.: Comparative performance analysis of clustering techniques in educational data mining. IADIS Int. J. Comput. Sci. Inf. Syst. **10**(2) (2015)
13. Martinez Abad, F., Chaparro Caso López, A.A.: Data-mining techniques in detecting factors linked to academic achievement. School Eff. School Improv. **28**(1), 39–55 (2017)
14. UAE Vision 2021. First-Rate Education System (2019). https://www.vision2021.ae/en/national-agenda-2021/list/first-rate-circle
15. Mullis, I.V.S., Martin, M.O. (eds.): TIMSS 2019 Assessment Frameworks. Boston College, TIMSS & PIRLS International Study Center (2017). http://timssandpirls.bc.edu/timss2019/frameworks/
16. Baradwaj, B.K., Pal, S.: Mining Educational Data to Analyze Students' Performance (2012). ArXiv:1201.3417
17. Ifenthaler, D., Yau, J.-K.: Utilizing learning analytics to support study success in higher education: a systematic review. Educ. Tech. Res. Dev. **68**(4), 1961–1990 (2020)
18. Kiray, S.A., Gok, B., Selman Bozkir, A.: Identifying the factors affecting science and mathematics achievement using data mining methods. J. Educ. Sci. Environ. Health **1**(1), 28–48 (2015)
19. Lee, J., Shute, V.J.: Personal and social-contextual factors in K–12 academic performance: an integrative perspective on student learning. Educ. Psychol. **45**(3), 185–202 (2010)
20. Akessa, G.M., Dhufera, A.G.: Factors that influences students' academic performance: a case of Rift Valley University, Jimma, Ethiopia. J. Educ. Pract. **6**(22), 55–63 (2015)

21. Kabakchieva, D.: Predicting student performance by using data mining methods for classification. Cybern. Inf. Technol. **13**(1), 61–72 (2013)
22. Lau, E.T., Sun, L., Yang, Q.: Modeling, prediction and classification of student academic performance using artificial neural networks. SN Appl. Sci. 1(9), 1–10 (2019)
23. Liem, G.A.D., Martin, A.J., Porter, A.L., Colmar, S.: Sociocultural antecedents of academic motivation and achievement: role of values and achievement motives in achievement goals and academic performance. Asian J. Soc. Psychol. **15**(1), 1–13 (2012)
24. Schumacher, P., Olinsky, A., Quinn, J., Smith, R.: A comparison of logistic regression, neural networks, and classification trees predicting success of actuarial students. J. Educ. Bus. **85**(5), 258–263 (2010)
25. Bahadır, E.: Using neural network and logistic regression analysis to predict prospective mathematics teachers' academic success upon entering graduate education. Kuram ve Uygulamada Egitim Bilimleri **16**(3), 943–964 (2016). https://doi.org/10.12738/estp.2016.3.0214
26. De Witte, K., Kortelainen, M.: What explains the performance of students in a heterogeneous environment? Conditional efficiency estimation with continuous and discrete environmental variables. Appl. Econ. **45**(17), 2401–2412 (2013)
27. Nath, S.R.: Factors influencing primary students' learning achievement in Bangladesh. Res. Educ. **88**(1), 50–63 (2012)
28. Mohtar, L.E., Halim, L., Samsudin, M.A., Ismail, M.E.: Non-cognitive factors influencing science achievement in Malaysia and Japan: an analysis of TIMSS 2015. EURASIA J. Math. Sci. Technol. Educ. **15**(4), 1697 (2019)
29. Pérez, P.M., Castejón Costa, J.-L., Corbi, R.G.: An explanatory model of academic achievement based on aptitudes, goal orientations, self-concept and learning strategies. Span. J. Psychol. **15**(1), 48–60 (2012)
30. Yoo, J.E., Rho, M.: TIMSS 2015 Korean student, teacher, and school predictor exploration and identification via random forests. SNU J. Educ. Res. **26**(4), 43–61 (2017). https://s-space.snu.ac.kr/bitstream/10371/168474/1/26(4)_03.pdf. Accessed 30 March 2022
31. Mohammadpour, E., Shekarchizadeh, A., Kalantarrashidi, S.A.: Multilevel modeling of science achievement in the TIMSS participating countries. J. Educ. Res. **108**(6), 449–464 (2015). https://doi.org/10.1080/00220671.2014.917254
32. Bernardo, A.B., Cordel, M.O., Lucas, R.I., Teves, J.M., Yap, S.A., Chua, U.C.: Using machine learning approaches to explore non-cognitive variables influencing reading proficiency in English among Filipino learners. Educ. Sci. **11**(10), 628 (2021). https://doi.org/10.3390/educsci11100628
33. Nadaf, A., Eliëns, S., & Miao, X.: Interpretable-machine-learning evidence for importance and optimum of learning time. Int. J. Inf. Educ. Technol. **11**(10), 444–449 (2021). https://doi.org/10.18178/ijiet.2021.11.10.1548
34. Lundberg, S., Lee, S.-I.: A unified approach to interpreting model predictions. Adv. Neural Inf. Process. Syst. **30**, 4765–4774 (2017)
35. Lundberg, S.M., Erion, G.G., Lee, S.-I.: Consistent Individualized Feature Attribution for Tree Ensembles (2019). ArXiv:180203888 Cs Stat
36. Lundberg, S.M., Erion, G., Chen, H., DeGrave, A., Prutkin, J.M., Nair, B., Katz, R., Himmelfarb, J., Bansal, N., Lee, S.-I.: From local explanations to global understanding with explainable AI for trees. Nat. Mach. Intell. **2** (2020). https://doi.org/10.1038/s42256-019-0138-9. https://par.nsf.gov/biblio/10167481
37. Fishbein, B., Foy, P., Yin, L.: TIMSS 2019 User Guide for the International Database, 2nd ed. Boston College, TIMSS & PIRLS International Study Center (2021). https://timssandpirls.bc.edu/timss2019/international-database/
38. Martin, M.O., von Davier, M., Mullis, I.V.: Methods and Procedures: TIMSS 2019 Technical Report. International Association for the Evaluation of Educational Achievement (2020)

39. CatBoost.: github.com/catboost/catboost (2020). [Online]. https://github.com/catboost/catboost
40. Hair, J.F., Ringle, C.M., Sarstedt, M.: PLS-SEM: indeed a silver bullet. J. Mark. Theory and Pract. **19**(2), 139–152 (2011). https://ssrn.com/abstract=1954735
41. Teo, T.W., Choy, B.H.: In: Tan, O.S., Low, E.L., Tay, E.G., Yan, Y.K. (eds.) Singapore Math and Science Education Innovation. ETLPPSIP, vol. 1, pp. 43–59. Springer, Singapore (2021). https://doi.org/10.1007/978-981-16-1357-9_3
42. OECD: Early learning matters, the international early learning and child well-being study (2018)
43. OECD: Better Skills, Better Jobs, Better Lives: A Strategic Approach to Education and Skills Policies for the United Arab Emirates (2015). https://www.oecd.org/education/A-Strategic-Approach-to-Education-and%20Skills-Policies-for-the-United-Arab-Emirates.pdf
44. City, E.A., Elmore, R.F., Fiarman, S.E., Teitel, L.: A Network Approach to Improving Teaching and Learning. Harvard Education Press, Cambridge (2009)
45. McCombs, B.L.: The role of the self-system in self-regulated learning. Contemp. Educ. Psychol. **11**, 314–332 (1986)

Educational Information Technology and E-Learning

Explainability in Automatic Short Answer Grading

Tim Schlippe[✉], Quintus Stierstorfer, Maurice ten Koppel, and Paul Libbrecht

IU International University of Applied Sciences, Erfurt, Germany
tim.schlippe@iu.org

Abstract. Massive open online courses and other online study opportunities are providing easier access to education for more and more people around the world. To cope with the large number of exams to be assessed in these courses, AI-driven automatic short answer grading can recommend teaching staff to assign points when evaluating free text answers, leading to faster and fairer grading. But what would be the best way to work with the AI? In this paper, we investigate and evaluate different methods for explainability in automatic short answer grading. Our survey of over 70 professors, lecturers and teachers with grading experience showed that displaying the predicted points together with matches between student answer and model answer is rated better than the other tested explainable AI (XAI) methods in the aspects *trust, informative content, speed, consistency and fairness, fun, comprehensibility, applicability, use in exam preparation*, and *in general*.

Keywords: Explainability · Explainable AI · XAI · Automatic short answer grading · AI in education

1 Introduction

Access to education is one of people's most important assets and ensuring inclusive and equitable quality education is goal 4 of United Nations' Sustainable Development Goals [1]. Distance learning in particular can create education in areas where there are no educational institutions or in times of a pandemic. There are more and more offers for distance learning worldwide and challenges like the physical absence of the teacher and the classmates or the lack of motivation of the students are countered with technical solutions like videoconferencing systems [2] and gamification of learning [3]. The research area "AI in Education" addresses the application and evaluation of Artificial Intelligence (AI) methods in the context of education and training [4–6]. One of the main focuses of this research is to analyze and improve teaching and learning processes. Many educational institutions—public and private—already conduct their courses and examinations online. This means that student examinations and their assessments are already available in digital, machine-readable form, offering a wide range of analysis options. An exam often consists of multiple choice and free text questions. Multiple choice questions can be made so that answers are unambiguous and have been easy to evaluate by machine for many years. However, the evaluation of free text answers—i.e., assigning quantitative feedback on the correctness of the student response in the form of a score possible within a certain point range—still required tedious manual work by the graders since it was a greater challenge for automatic systems. Fortunately, automatic short answer grading (ASAG) is improving and in some cases has already reached the point where teaching staff could use it for fairer and faster grading [7]. Since the results in terms of scores are not yet perfect and graders want to understand ASAG systems' decisions, the question arises: What is the best way to make the decisions of ASAG systems explainable to human graders? The desire for explainability is also demonstrated by the feedback from 71 professors, lecturers, and teachers, as shown in Fig. 1. A clear majority of participants strongly agrees that it is important for them to understand how an AI reaches its expected scoring (4.42 on average) and their confidence in an AI grading support increases when it explains itself (4.18 on average).

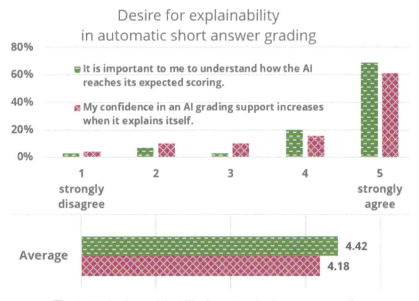

Fig. 1. Desire for explainability in automatic short answer grading.

The field of explainable AI (XAI) aims to provide solutions for the need of transparency. XAI can be described as research direction that aims to create human interpretable AI [8]. In this paper, we investigate and evaluate different methods for explainability in ASAG. Thus, we provide insight into the perceived usefulness of different XAI methods for ASAG. For the evaluation of the XAI methods we asked over 70 professors, lecturers, and teachers to rate different aspects.

In the next section, we will present the latest approaches of other researchers for ASAG and explainability. Section 3 will demonstrate our investigated methods for explainability in ASAG. Section 4 will describe the experimental setup for our user study. The study and the results are outlined in Sect. 5. We will conclude our work and suggest further steps in Sect. 6.

2 Related Work

In this section we will present related work in the areas of explainability and ASAG.

2.1 Explainability

Before the deep learning era, data scientists crafted predictive models by manually inspecting the data and constructing models based on the insights. With deep learning, the best practice is now to let the algorithm figure out itself which parts of the data are useful [9]. Typically, modern deep learning models are created with hundreds of features using gigabyte-sized data sets. Verification of these models is usually done by calculating accuracy: The goal is that a model makes as few errors as possible ignoring the reasons why a prediction is made. However, experts are now beginning to understand that this metric alone is not enough [10]. We are beginning to ask questions such as: Is a model robust to small variations in data? How does the model behave when atypical data is input? Is a model safe to use or could data be extracted? Does the model respect privacy of individuals? Is it fair and non-discriminatory?

The field of XAI aims to provide solutions for these transparency needs. Although the field lacks a clear definition, it can be best described as a research direction that aims to create human interpretable or explainable AI [1]. The human interpretation part of the explanation is critical: If it cannot be understood or applied by a human, it becomes meaningless. The field of XAI is therefore considered to be a multi-disciplinary field in which data scientists, AI engineers, human scientists, and human-computer interaction specialists work together to create technical sound and human interpretable explanations. XAI has a similarly long history like the field of AI itself. However, it first started to boom after the deep learning revolution in 2012 [11]. Good overviews of the field are provided by [8, 12–15].

On the most basic level, two directions can be identified within this field: (i) Methods that aims to make models intrinsically interpretable and (ii) methods that aim to provide transparency for black box models (post-hoc methods). The latter is the vast dominating approach in the field. Black box models are models that "are created directly from data by an algorithm, meaning that humans, even those who design them, cannot understand how variables are being combined to make predictions" [16]. To deal with those black box models, ongoing research is the creation of additional predictive models that can meet the accuracy of black box models while being intrinsically interpretable.

Since the state-of-the-art ASAG models are mostly based on transformer models, they are also black box models. Therefore, our XAI methods are not directly based on an interpretation of the actual existing ASAG model but use separate models for explainability.

2.2 Automatic Short Answer Grading

The field of ASAG is becoming more relevant since many educational institutions—public and private—already conduct their courses and examinations online [7, 17]. A good overview of approaches in ASAG before the deep learning era is given in [18]. [19] and [17] investigate and compare state-of-the-art deep learning techniques for ASAG.

[19] demonstrate that systems based on BERT performed best for English and German and that their multilingual RoBERTa model [20] shows a stronger generalization across languages on English and German. [7] extended ASAG to 26 languages and use the smaller M-BERT [21] model to conduct a larger study concerning the cross-lingual transfer. With Mean Absolute Errors between 0.41 and 0.72 points out of 5 points they report that their best models have even less discrepancy than 2 graders, which is 0.75 points. This shows that the performances are now good enough to be used as a support for scoring answers to open exam questions—provided that the prediction of AI is presented in a good way to the graders.

A first work towards explainability in ASAG is described by [22]. However, their focus is a comparison of different corpora and feature attribution techniques to "attribute" words in the student answer, which should then make the decision of the ASAG system more explainable. While the focus of [22] is on the technical implementation of their attribution values with transformer-based models, our work is more general as it does not depend on specific models, evaluates the perception of a broader range of XAI methods and has a strong focus on feedback of the teaching staff through detailed evaluation of 9 aspects as described in Sects. 4 and 5. We can imagine that our results can be used complementary to the results of [22].

3 Methods for Explainability in Automatic Short Answer Grading

The goal of ASAG is to enable (semi) automatic, fair, and consistent grading at a high productivity. We explored if and how explanations might be beneficial to this use case. Figure 2 visualizes with an example the pipeline of our systems for AI-driven grading support. The system consists of 1 ASAG model for point prediction and 1 model for explainability. The ASAG model always processes the student answer and the model answer[1] for the prediction. Depending on the XAI model, only the student answer or the student answer plus the model answer is input. In our experiments, we exchanged the XAI method so that its output—here *Matching Positions*—is different for each method.

In our investigation we examined common XAI method classes described in the literature which could be adapted to the ASAG task—even if they have only been used for Computer Vision and not yet for Natural Language Processing (NLP). Table 1 demonstrates 3 prevalent common XAI method classes which are appropriate for this task and describes their characteristics.

From the remaining XAI method classes, we created more specific methods useful for the ASAG task. Table 2 lists our 5 created XAI methods along with the method classes on which their creation is based. Only *Predicted Points* is not based on a method class, since only the number of points is displayed.

In the next sections, our developed ASAG-specific XAI methods are described in more detail.

3.1 Predicted Points

As illustrated in Fig. 3, displaying only the predicted points is the simplest of the evaluated methods (*Predicted Points*). No additional XAI model is used.

---
[1] Also called *sample answer* or *sample response* in literature.

Explainability in Automatic Short Answer Grading   73

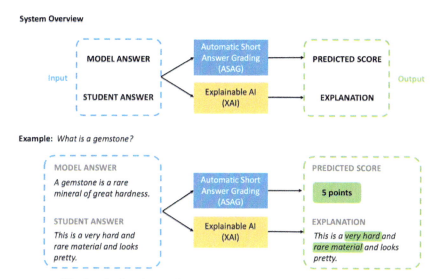

Fig. 2. Pipeline for grading with point prediction and explainability.

Table 1. XAI method classes suitable for the generation of XAI methods specific for ASAG.

XAI method class	Description
Confidence score	Certainty of a model's prediction is made interpretable and inspectable [23]
Word highlighting	Words are color marked to indicate their relevance towards the classification [24]
Concept activation	High level human concepts are used to explain a classification [25]

Table 2. XAI methods specific for ASAG.

XAI method for ASAG	XAI method class
Predicted points	–
Predicted points with confidence scores	*Confidence score*
Predicted points with confidence scores and similar answers	*Confidence score*
Predicted points with relevance of words in the answer	*Word highlighting*
Predicted points with matching positions	*Concept activation*

3.2 Predicted Points with Confidence Scores

As demonstrated in Fig. 4 (*Predicted Points with Confidence Scores*), interpretable confidence scores put the predicted score in context of past performance. This confidence score is computed by considering similar past cases where the AI and human collaborated, and

Fig. 3. XAI Method: Predicted Points.

hence the system has feedback on the accuracy of its predictions. The interpretable confidence score is essentially a single percentage between 0% and 100%, with additional information on how many similar answers it is computed.

Fig. 4. XAI Method: Predicted Points with Confidence Scores.

3.3 Predicted Points with Confidence Scores and Similar Answers

Interpretable confidence scores can be extended by providing examples of answers that are similar as shown in Fig. 5 (*Predicted Points with Confidence Scores and Similar Answers*). Hence, the graders do not only get one confidence score but also see examples of answers that were rated equally. This concept with comparable scored answers could also help new graders get into the grading process more quickly as well as help students relate their answer to other answers and deduce why their own answer is incorrect or correct.

3.4 Predicted Points with Relevance of Words in the Answer

The standard method for explainability in the context of NLP is word highlighting [14, 24]. Typically, every word is marked to indicate its relevance for the prediction. However, this can be confusing and hard to interpret. Instead in this method a threshold of relevance makes sure that only a subset of the most important words is marked as indicated in Fig. 6 (*Predicted Points with Relevance of Words in the Answer*). The idea is that this leads to a more efficient interpretation of information.

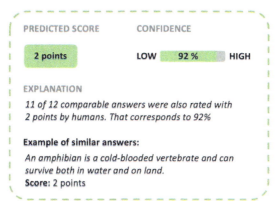

Fig. 5. XAI Method: Predicted Points with Confidence Scores and Similar Answers.

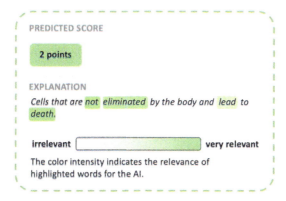

Fig. 6. XAI Method: Predicted Points with Relevance of Words in the Answer.

3.5 Predicted Points with Matching Positions

The method of concept attribution builds on the idea that a good explanation relates to understandable human concepts [25]. In its application for computer vision, not pixels themselves are highlighted, but instead an image is analyzed on the availability of a human understandable concept, e.g., the presence of medical condition. This method has not yet been demonstrated in NLP.

To transfer this idea to ASAG, we propose that the correct parts of the model answer are highlighted within the student answer (*Predicted Points with Matching Positions*) as shown in Fig. 7. In essence, the model answer is already a human understandable concept which an explanation should ideally relate to. We believe that this method leads to an efficient interpretation of information which should be useful to both the grader and the student.

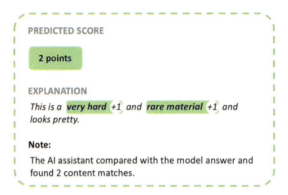

Fig. 7. XAI Method: Predicted Points with Matching Positions.

4 Experimental Setup

As mentioned before, we initially performed an analysis using XAI method classes that have been successfully proven for AI applications. The goal was to find those methods that are most promising in terms of use for graders' support. These 5 most promising methods were then evaluated in a survey by graders. Due to the appropriate range of functions, the good display of our images with the XAI methods and its platform independence, we conducted the survey with Google Forms[2].

As demonstrated in Fig. 8, the participants were asked to take the role of a teacher and evaluate various student answers given a model answer, a score predicted by the AI, and an explanation for the predicted score. In their role as graders, participants could assign 0, 1, or 2 points, with the highest possible score being 2 points. The type of explanation changed as the survey progressed. Before a new XAI method was displayed, questions were asked about the XAI method regarding the aspects *trust, informative content, speed, consistency and fairness, fun, comprehensibility, applicability, use in exam preparation,* and *in general*. The participants evaluated the questions regarding to the aspects with a score. The score range follows the rules of a forced choice Likert scale, which ranges from (1) *strongly disagree* to (5) *strongly agree*. Additionally, we evaluated the influence of the AI assistant's predictions on the graders' decisions by showing participants 2 correct and 1 incorrect point predictions of the AI assistant for each XAI method, i.e., one-third of the AI assistant's point prediction was incorrect.

71 participants (36 female, 35 male) filled out our questionnaire. The participants of our user study were professors, lecturers, and teachers between 24 and 65 years old who participated free of charge. Most are employed at our university. But some are professors and lecturers at other universities or teachers in schools. We appreciate these distributions as it was important to us to get feedback from different people.

---

[2] https://docs.google.com/forms.

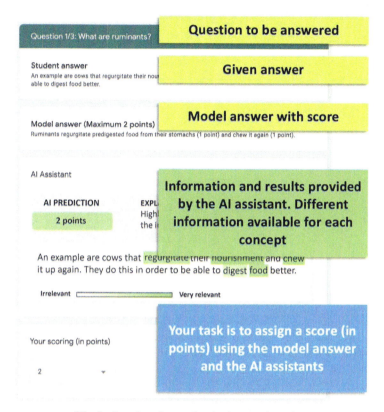

Fig. 8. Interface for grading in the questionnaire.

5 Experiments and Results

In this section, we will describe the results of our study in which we examined the XAI methods with regard to the aspects *trust, informative content, speed, consistency and fairness, fun, comprehensibility, applicability, use in exam preparation*, and *in general*. In addition, we investigated how strong the influence of the XAI method is on the grader's score—when the AI's prediction is correct and when it is not.

5.1 Trust

We asked the participants in our questionnaire if they based their evaluation on the AI assistant. The goal was to find out how much *trust* the participants have on the AI assistant depending on the XAI method. Figure 9 illustrates the feedback on the trust. On average, the highest trust is on *Points+matching positions* (3.30), followed by *Confidence+similar answers* (2.94), *Confidence* (2.80), *Relevance of words* (2.71), and *Points* (2.46).

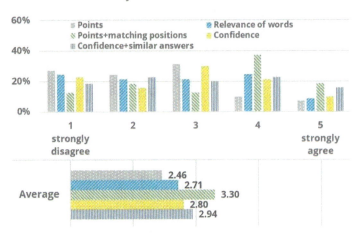

Fig. 9. Trust.

5.2 Informative Content

Figure 10 illustrates our evaluation in relation to the *informative content* of the suggested XAI methods. This time the averages are all above 3.00: *Points+matching positions* was rated on average with 3.57, *Confidence+similar answers* with 3.11, *Relevance of words* with 3.06, *Confidence* with 2.77, and *Points* with 2.60.

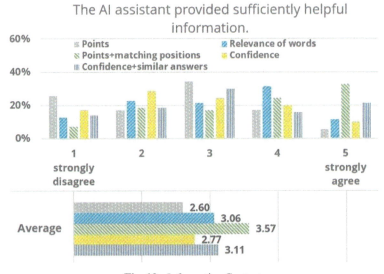

Fig. 10. Informative Content.

5.3 Speed

As illustrated in Fig. 11, in the category *speed* with 3.70 *Points+matching positions* is again the best rated method. In comparison, the question if the AI assistant could help evaluate faster was rated only with an average score of 3.29 with *Relevance of words*, 3.17 with *Points*, 3.04 with *Confidence+similar answers*, and 3.03 with *Confidence*.

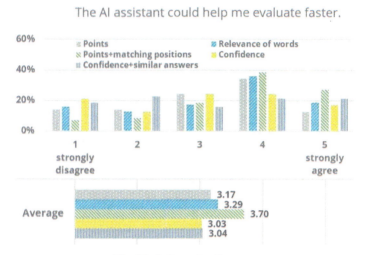

Fig. 11. Informative Content.

5.4 Consistency and Fairness

Then we asked the participants in our questionnaire if the AI assistant could help evaluate more *consistently and fairly*. The background to this is that graders do not always agree on the allocation of points and are also influenced in their grading by external factors that do not directly relate to the quality of the student answer [26]. The results are demonstrated in Fig. 12. This time, the averages are closer together: The highest value remains at *Points+matching positions* (3.44), followed by *Confidence+similar answers* (3.31), *Relevance of words* (3.24), *Points* (3.21), and *Confidence* (3.10).

5.5 Fun

We asked the participants in our questionnaire if the rating with the AI assistant would be *fun* using the selected XAI methods. The results are shown in Fig. 13. We see that on average, the highest value is again achieved by *Points+matching positions* (3.76), this time followed by *Relevance of words* and *Points* (3.59). Then comes *Confidence+similar answers* (3.26) and *Confidence* (3.16).

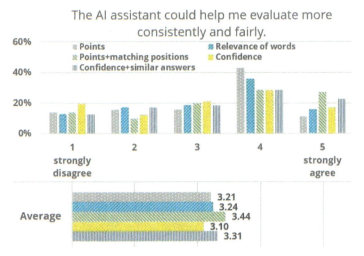

Fig. 12. Consistency and Fairness.

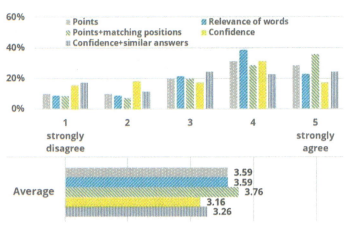

Fig. 13. Fun.

5.6 Comprehensibility

Another goal of our survey was to find out how comprehensible the XAI methods are. Therefore, our participants were asked if they were able to verify the recommendation of the AI assistant. Figure 14 illustrates the evaluation with regards to *comprehensibility*. Here, *Points+matching positions* (3.83), *Relevance of words* (3.76), and *Points* (3.61) are ahead on average. *Confidence+similar answers* (3.34) and *Confidence* (2.89) seem to be more difficult to understand on average.

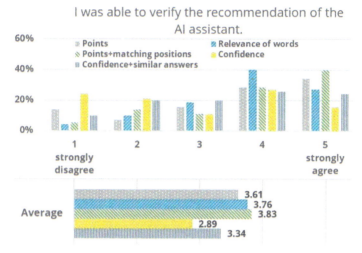

Fig. 14. Comprehensibility.

5.7 Applicability

The distribution in averages that we see in comprehensibility is also seen in *applicability* as shown in Fig. 15: The highest value remains at *Points+matching positions* (3.79), closely followed by *Relevance of words* (3.67), and *Points* (3.59). *Confidence+similar answers* (3.31) and *Confidence* (3.11) make up the tail.

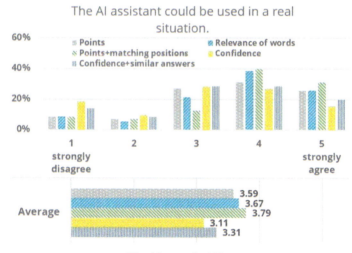

Fig. 15. Applicability.

5.8 Use in Exam Preparation

While the focus of the questions so far was on the use of XAI methods for the support of graders, XAI methods can also help learners prepare for an exam. Consequently, we asked our participants for each XAI method if they think that it is useful for learners as well. Again, *Points+matching positions* (3.41), *Relevance of words* (3.39) and *Points* (3.33) are close to each other. Again *Confidence+similar answers* (3.14) and *Confidence* (2.96) bring up the rear (Fig. 16).

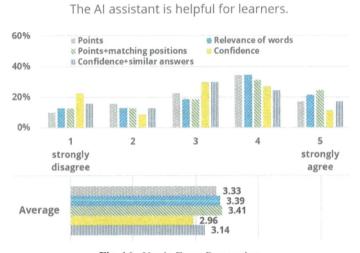

Fig. 16. Use in Exam Preparation.

5.9 General Evaluation

The last aspect we asked about was how the XAI methods are evaluated *in general*. As illustrated in Fig. 17, the trends are comparable as in the other aspects: With 3.94 on average, *Points+matching positions* is rated as good. Then comes *Relevance of words* (3.57) with 10% less, followed by *Confidence+similar answers* (2.97), *Confidence* (2.80), and *Points* (2.54).

5.10 Influence

Finally, we investigated how strong the *influence* of the XAI method is on the grader's score—when the AI's prediction is correct and when it is not. We evaluated the influence by showing participants 2 correct and 1 incorrect point predictions of the AI assistant for each XAI method, i.e., one-third of the AI assistant's point prediction was incorrect.

Figure 18 visualizes the percentage of correct scored student answers by the graders and the average deviations from the correct score in the case of a correct point prediction by the AI assistant and in the case of an incorrect point prediction by the AI assistant for

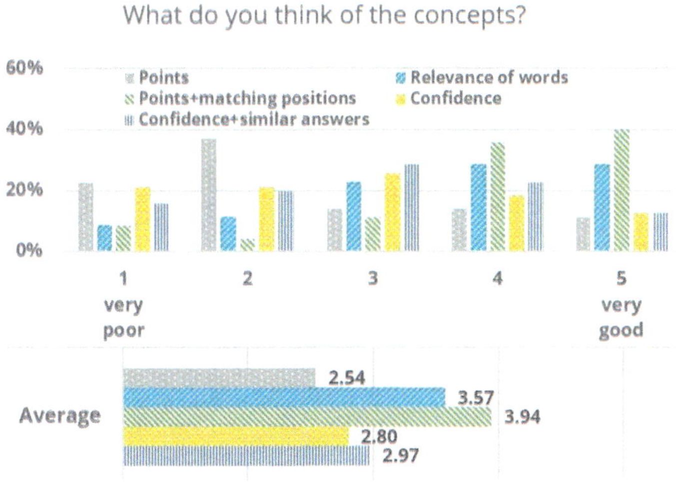

Fig. 17. Use in Exam Preparation.

all tested XAI methods. With a maximum score to be obtained per question of 2 points, the deviations of the graders range between 0.20 points (*Points+matching positions and relevance of words*) and 0.28 points (*Confidence*), which is only between 10 and 14%. For comparison, in the literature deviations of 15% between 2 graders are reported [17, 26], which shows that the tested XAI methods have no bad influence on the grading process. While in the case of a correct point prediction an average of 75–98% of the assessments achieved the same score as the assigned reference score, in the case of an incorrect point prediction an average of between 45 and 70% achieved the same score as the assigned reference score. For *points+matching positions*, which performed best in the previous questions, 89% graded correctly in the case of correct point prediction and 63% in the case of incorrect point prediction. One must keep in mind here that in our study, one-third of the AI assistant's point prediction was incorrect.

6 Conclusion

In this paper, we have investigated and evaluated different methods for explainability in ASAG. Our survey of over 70 professors, lecturers and teachers with grading experience showed that a clear majority of participants strongly agrees that it is important for them to understand how the AI reaches its expected scoring and their confidence in an AI grading support increases when it explains itself. Displaying the predicted points together with matches between student answer and model answer is rated better than the other tested XAI methods. Participants were asked if they agreed that the displayed XAI method helps for the aspects *trust, informative content, speed, consistency and fairness, fun, comprehensibility, applicability, use in exam preparation,* and *in general*.

Table 3 summarizes the average Likert scores for each evaluated aspect of the best method (*Points+matching positions*). Here the positive tendency for this method is shown as the scores are between 3 and 4, where 1 means *completely disagree* and 5 means

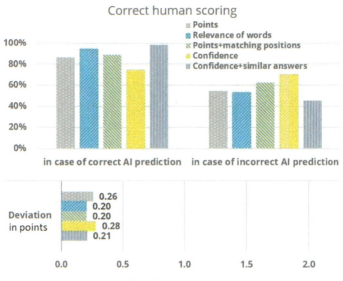

Fig. 18. Influence.

completely agree. In addition, the relative improvement compared to the second-best method in each of the aspects is demonstrated. The statistical significance was tested with the type I error $p = 2.5\%$ with a Student's *t*-test for paired samples with $n = 71$.

Table 3. Average Likert scores for *points+matching positions* over the evaluated aspects and rel. improvement to 2nd-best XAI method.

Aspect	Ø Likert score	Improvement over 2nd-best XAI method
trust	3.30	+12.2%* (*Confidence + similar answers*)
informative content	3.57	+14.8%* (*Confidence + similar answers*)
speed	3.70	+12.5%* (*Relevance of words*)
consistency & fairness	3.44	+3.9% (*Confidence + similar answers*)
fun	3.76	+4.7%* (*Relevance of words/Points*)
comprehensibility	3.83	+4.5% (*Relevance of words*)
applicability	3.79	+3.3%* (*Relevance of words*)
use in exam preparation	3.41	+0.6% (*Relevance of words*)
in general	3.94	+10.4%* (*Relevance of words*)

* statistically significant

Additionally, we investigated how strong the influence of the XAI method is on the grader's score—when the AI's prediction is correct and when it is not. The deviations of the graders from the actual points ranged between 10% and 14%. For comparison, in the literature deviations of 15% between 2 graders are reported [7, 27], which shows that the tested XAI methods have only little influence on the overall grading process.

7 Future Work

Our goal was to survey a large representative group of teaching staff consisting of professors, lecturers and teachers with exam questions, student answers and model answers that are understandable for all participants. Therefore, we did not make an analysis of the individual lecturers' experience, subjects taught, performance at different difficulty levels, etc. This could be investigated in more detail in a future analysis.

Due to the very high performance of point prediction and the good results of the XAI methods in our survey, we plan to use ASAG together with the best XAI method at our university. In addition to grading, ASAG can also be used for exam preparation with an app or in online learning [28]. Consequently, future work may include to analyze the use of our XAI methods in interactive training programs to prepare students optimally for exams. Since in our study we considered the ASAG model as a black box model and produced explainability with another model, a graders' support by the direct interpretation of the complex ASAG models could be also investigated.

References

1. United Nations: Sustainable development goals: 17 goals to transform our world (2021). https://www.un.org/sustainabledevelopment/sustainable-development-goals
2. Correia, A.P., Liu, C., Xu, F.: Evaluating videoconferencing systems for the quality of the educational experience. Distance Educ. **41**(4), 429–452 (2020). https://doi.org/10.1080/01587919.2020.1821607
3. Koravuna, S., Surepally, U.K.: Educational gamification and artificial intelligence for promoting digital literacy. Association for Computing Machinery, New York, NY, USA (2020)
4. Chen, L., Chen, P., Lin, Z.: Artificial intelligence in education: A review. IEEE Access **8**, 75264–75278 (2020). https://doi.org/10.1109/ACCESS.2020.2988510
5. Heffernan, N.T., Heffernan, C.L.: The ASSISTments ecosystem: Building a platform that brings scientists and teachers together for minimally invasive research on human learning and teaching. Int. J. Artif. Intell. Educ. **24**(4), 470–497 (2014). https://doi.org/10.1007/s40593-014-0024-x
6. Libbrecht, P., Declerck, T., Schlippe, T., Mandl, T., Schiffner, D.: NLP for student and teacher: Concept for an AI based information literacy tutoring system. In: The 29th ACM International Conference on Information and Knowledge Management (CIKM2020). Galway, Ireland (2020)
7. Schlippe, T., Sawatzki, J.: Cross-lingual automatic short answer grading. In: Proceedings of the 2nd International Conference on Artificial Intelligence in Education Technology (AIET 2021). Wuhan, China (2021)
8. Adadi, A., Berrada, M.: Peeking inside the black-box: A survey on explainable artificial intelligence (XAI). IEEE Access **6**, 52138–52160 (2018). https://doi.org/10.1109/ACCESS.2018.2870052(2018)
9. Ng, A.: Machine learning yearning. Online draft. https://github.com/ajaymache/machine-learning-yearning (2017)

10. Doshi-Velez, F., Kim, B.: Towards a rigorous science of interpretable machine learning (2017). arXiv:1702.08608
11. Hansen, L.K., Rieger, L.: Interpretability in intelligent systems—a new concept? In: Samek, W., Montavon, G., Vedaldi, A., Hansen, L.K., Müller, K.-R. (eds.) Explainable AI: Interpreting, Explaining and Visualizing Deep Learning. LNCS (LNAI), vol. 11700, pp. 41–49. Springer, Cham (2019). https://doi.org/10.1007/978-3-030-28954-6_3
12. Bodria, F., Giannotti, F., Guidotti, R., Naretto, F., Pedreschi, D., Rinzivillo, S.: Benchmarking and survey of explanation methods for black box models (2021). arXiv:2102.13076
13. Carvalho, D.V., Pereira, E.M., Cardoso, J.S.: Machine learning interpretability: A survey on methods and metrics. Electronics **8**(8) (2019). https://doi.org/10.3390/electronics8080832
14. Danilevsky, M., Qian, K., Aharonov, R., Katsis, Y., Kawas, B., Sen, P.: A survey of the state of explainable AI for natural language processing (2020). arXiv:2010.00711
15. Samek, W., Montavon, G., Lapuschkin, S., Anders, C.J., Müller, K.R.: Explaining deep neural networks and beyond: A review of methods and applications. Proc. IEEE **109**(3), 247–278 (2021). https://doi.org/10.1109/JPROC.2021.3060483
16. Rudin, C., Radin, J.: Why are we using black box models in AI when we don't need to? A lesson from an explainable AI competition. Harvard Data Science Issue 1.2 (2019). https://doi.org/10.1162/99608f92.5a8a3a3d
17. Sawatzki, J., Schlippe, T., Benner-Wickner, M.: Deep learning techniques for automatic short answer grading: Predicting scores for English and German answers. In: Proceedings of The 2nd International Conference on Artificial Intelligence in Education Technology (AIET 2021). Wuhan, China (2021)
18. Burrows, S., Gurevych, I., Stein, B.: The eras and trends of automatic short answer grading. Int. J. Artif. Intell. Educ. **25**(1), 60–117 (2014). https://doi.org/10.1007/s40593-014-0026-8
19. Camus, L., Filighera, A.: Investigating transformers for automatic short answer grading. In: Bittencourt, I.I., Cukurova, M., Muldner, K., Luckin, R., Millán, E. (eds.) AIED 2020. LNCS (LNAI), vol. 12164, pp. 43–48. Springer, Cham (2020). https://doi.org/10.1007/978-3-030-52240-7_8
20. Liu, Y., Ott, M., Goyal, N., Du, J., Joshi, M., Chen, D., Levy, O., Lewis, M., Zettlemoyer, L., Stoyanov, V.: RoBERTa: A robustly optimized BERT pretraining approach. CoRR (2019). arXiv:1907.11692
21. Pires, T., Schlinger, E., Garrette, D.: How multilingual is multilingual BERT? In: Proceedings of the 57th Annual Meeting of the Association for Computational Linguistics. Association for Computational Linguistics, Florence, Italy, pp. 4996–5001 (2019). https://doi.org/10.18653/v1/P19-1493
22. Poulton, A., Eliens, S.: Explaining transformer-based models for automatic short answer grading. In: Proceedings of the 5th International Conference on Digital Technology in Education (ICDTE 2021). Association for Computing Machinery, New York, NY, USA, pp. 110–116 (2021). https://doi.org/10.1145/3488466.3488479
23. van der Waa, J., Schoonderwoerd, T., van Diggelen, J., Neerincx, M.: Interpretable confidence measures for decision support systems. Int. J. Hum.-Comput. Stud. 144 (2020). https://doi.org/10.1016/j.ijhcs.2020.102493
24. Ribeiro, M.T., Singh, S., Guestrin, C.: "Why should I trust you?": Explaining the predictions of any classifier. In: Proceedings of the 22nd ACM SIGKDD International Conference on Knowledge Discovery and Data Mining (San Francisco, California, USA) (KDD '16). Association for Computing Machinery, New York, NY, USA, pp. 1135–1144 (2016). https://doi.org/10.1145/2939672.2939778
25. Kim, B., Wattenberg, M., Gilmer, J., Cai, C.J., Wexler, J., Viégas, F., Sayres, R.: Interpretability beyond feature attribution: Quantitative testing with concept activation vectors (TCAV). In: ICML 2018

26. Hanna, R.N., Linden, L.L.: Discrimination in grading. Am. Econ. J. Econ. Policy **4**(4), 146–168 (2012). http://www.jstor.org/stable/23358248
27. Mohler, M., Bunescu, R., Mihalcea, R.: Learning to grade short answer questions using semantic similarity measures and dependency graph alignments. In: Proceedings of the 49th Annual Meeting of the Association for Computational Linguistics: Human Language Technologies. Association for Computational Linguistics, Portland, Oregon, USA, pp. 752–762 (2011)
28. Schlippe, T., Sawatzki, J.: AI-based multilingual interactive exam preparation. In: Guralnick, D., Auer, M.E., Poce, A. (eds.) TLIC 2021. LNNS, vol. 349, pp. 396–408. Springer, Cham (2022). https://doi.org/10.1007/978-3-030-90677-1_38

The Framework Design of Intelligent Assessment Tasks Recommendation System for Personalized Learning

Qihang Cai and Lei Niu[✉]

Faculty of Artificial Intelligence in Education, Central China Normal University,
Wuhan 430079, China
caiqihang19981003@126.com, lniu@ccnu.edu.cn

Abstract. In teaching, assessment tasks are often used as an important way to evaluate students' learning abilities. In traditional education, to design an assessment task, e.g., an assignment, teachers are often required to manually design by themselves. It is usually a challenging task to design high-quality assignments, especially for less-experienced teachers. In addition, students often have different learning abilities and it is often difficult and unreasonable to evaluate students' learning abilities using only the same assignments. Moreover, to design an assignment with decent quality, teachers have to consider the knowledge items to be covered and the difficulty of the assignment. Therefore, it is worthwhile to do the research of automatically providing students with high-quality assessment tasks, taking into account the coverage of knowledge items and the appropriate difficulty of the designed assessment tasks, i.e., proposing an approach of personalized assessment task recommendation systems. The current literature on personalized assessment task recommendations shows that the recommendation process does not take into account the difficulty of the designed questions and students' knowledge. To overcome the limitations of the current related work, this paper proposes a framework design of intelligent assessment tasks recommendation by considering several aspects, such as students' learning ability, mastery of student knowledge, the difficulty of assessment tasks, and students' forgetting characteristics. The proposed framework design consists of two components: data processing and personalized assessment task generation. The data processing component is designed to proceed with data, e.g., generating the initial question bank, analyses on the question bank generated, auto-generation of the assessment task, and the result collection of auto-correcting on the designed assessment task. Besides, the forgetting characteristics of students are also considered in this study for the intelligent assessment tasks recommendation framework design.

Keywords: Personalized learning · Recommendation system · Difficulty calculation of assessment task · Students' learning abilities evaluation · Framework design

1 Introduction

Teaching refers to planned purposeful activities organized by teachers for students. Usually, teachers help students achieve effective learning, acquire knowledge and promote their development. In teaching, it is very important to assess students' learning performance.

In students' learning performance evaluation, assessment tasks are independent learning activities undertaken by students in their studies, and it helps evaluate both students' learning quality and lecturers' teaching quality. In traditional education, to produce an appropriate assessment task (e.g., an assignment) for students, teachers have to create an assignment by themselves, considering the knowledge items to be covered in the assignment as well as the difficulty of the overall assignment based on their teaching experiences.

However, teachers with little experience, e.g., fresh lecturers, might not be able to create an assignment for students with decent quality, which results in the inaccurate outcome of evaluating students' learning performance through designed assignment. Therefore, providing decent assessment tasks for students considering knowledge items covered and appropriate difficulty is worth studying [1].

Besides, considering independent students with different levels of learning abilities, it is usually very difficult to evaluate all students with only one single assignment constituted by the same questions. That is to say, traditional education cannot well provide personalized assignment recommendations. Therefore, to better overcome such a problem in traditional education, proposing an intelligent assessment tasks recommendation framework for independent students considering their different personalities is greatly challenging.

In current literature, Hussain et al. [2] use artificial neural networks, decision trees, logistic regression, and Support Vector Machines (SVM) to predict students' performance in subsequent sessions of a digital design course to identify students who needed additional help. Benhamdi et al. [3] provide students with the best learning materials based on their interests, background knowledge, and memory capabilities through a new recommendation method NPR_eL, based on collaboration and content filtering. Karga and Satratzemi [4] use a similarity matrix, combining collaborative filtering and content-based recommendation to help teachers redesign the learning design process according to students' needs and preferences. However, in the above articles, the recommendation process does not take into account the difficulty of the questions and the knowledge items covered in the recommended assignments.

To overcome the limitation of current related work and resolve the problem of the assessment task recommendation approach for personalized education, this paper proposes an assessment task recommendation framework for personalized education by considering various aspects, i.e., student's learning ability, knowledge items covered, the difficulty level of the whole assessment task, the student's forgetting characteristic, etc.

In detail, this paper first introduces the whole recommendation framework design, then explains the details of the components contained in the proposed recommendation framework. The framework design of the intelligent personalized assessment task recommendation system consists of two components: data processing and personalized assessment tasks generation.

The data processing component is designed to process data, which helps to prepare question banks, such as question bank preparation, and question analysis. Also, it has the functions of intelligent paper generation and automatic correcting. To create assessment tasks for each student, an intelligent personalized assessment tasks recommendation component is designed. The recommendation process is based on the student's learning ability. In the component, we 1) consider student's learning ability, 2) the most appropriate level of difficulty for the assessment tasks, 3) each student's mastery of knowledge items, and 4) provide personalized assessment tasks for each student based on appropriate difficulty and mastery of knowledge items. Besides, the intelligent personalized assessment task recommendation framework also takes into account the level of forgetfulness of students and provides assessment tasks for students on forgotten knowledge items.

This paper is organized as follows. Section 2 provides related work. Section 3 introduces the whole recommendation framework design. Section 4 provides details of each component contained in the proposed recommendation framework. Section 5 concludes the paper.

2 Related Work

There are usually three approaches to recommend assessment tasks: Collaborative Filtering (CF), Content-Based recommendation (CB), and hybrid recommendation [5].

CB algorithms are based on historical data to determine students' preferences to recommend assignments [5]. Chen et al. [6] apply Skill Groups (SGs) according to the relevance of knowledge. For each SG, concept maps and blue-red trees are generated to describe the connections between learning courses, which are used to optimize the learning path sequence for students. Similarly, Kolekar et al. [7] develop a rule-based adaptive user interface that identifies the student's learning style to plan the student's learning path and deliver learning content. Shu et al. [8] use historical student data and a latent factor model to calculate learning characteristics and preferences, and use the CB algorithm, based on CNN, to predict the learning material. CB algorithm relies on historical data and cannot recommend new learning materials, which may reduce students' motivation.

Collaborative filtering recommendations are based on user behavior and ratings, recommending similar assignments and exploring what students might be interested in learning [5]. Based on feature similarities of students, Zhou et al. [9] train a long short-term memory model to predict students' learning paths and performances to recommend assessment tasks. Yera and Martínez [10] address the problem of information overload during the use of the programming online judge by using a collaborative filtering recommendation method to find programming problems suitable for students' programming skills. Dwivedi et al. [11] design an effective learning path for students based on their learning style and knowledge level through a variable-length genetic algorithm. Limitations of the CF method are that when data is scarce, recommendations become inaccurate and can create cold start problems, and the method is not very scalable, so in general, CF cannot be used as part of an efficient real-time recommendation system [5].

Hybrid recommendation combines the characteristics of both CB and CF by combining the predictions of both models into one, by adding the information obtained from CB to CF, or by obtaining a final recommendation through a weighted average or combined ranking of CB and CF [5]. This technique overcomes the drawbacks of both CB and CF and it also appears to be more accurate in cases where there are few ratings. The intelligent personalized assessment task recommendation system is based on the student's academic performance to recommend assignments for the student, as well as historical data to recommend assignments related to forgotten knowledge items.

3 Assessment Tasks Recommendation Framework Design

The framework design of the intelligent personalized assessment tasks recommendation is shown in Fig. 1, the framework design consists of two components: data processing and personalized assessment task generation.

3.1 Data Processing Component

The data processing component is designed to proceed with data, which contains three functions: question bank preparation, intelligent auto-generating test paper, and intelligent correcting.

Question bank preparation: This function concludes in two parts, question bank generation, which is used to collect questions, and question analysis, which can identify the knowledge items and calculate the difficulty of questions.

Intelligent auto-generating test paper: This function generates initial assessment tasks for students through setting expected difficulty and knowledge items.

Intelligent correcting: This function is capable of recognizing handwriting. Based on the identified information, this function corrects results of objective questions and corrects results and steps of subjective questions.

3.2 Personalized Assessment Task Generation Component

The personalized assessment task generation component divides into the four functions: evaluation of students' learning abilities, calculation of the appropriate level of difficulty and mastery of knowledge items, generation of personalized assessment tasks, and forgetfulness-based assessment tasks recommendations.

1. *Evaluation of students' learning abilities*: To acquire personalized assessment tasks recommendation, we need to evaluate the learning abilities of each student. Initial assessment tasks, which are generated by an intelligent auto-generating test paper function, are the sources of data. This function can get the difficulty of the assessment tasks from the question bank and according to the intelligent correcting function, the function obtains the correctness of the answers for students. According to the difficulty of assessment tasks and the correctness of answers, students' learning abilities are evaluated.

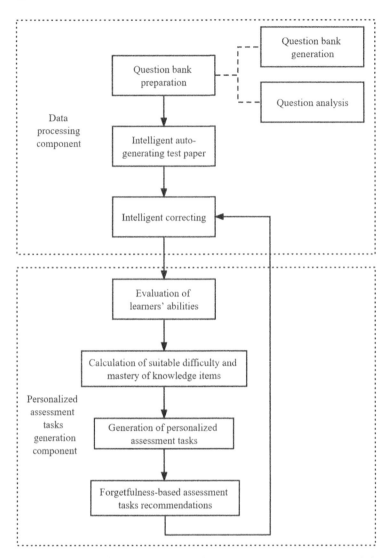

Fig. 1. Framework design of intelligent personalized assessment tasks recommendation system

2. *Calculation of appropriate difficulty and mastery of knowledge items*: The first is the calculation of the appropriate difficulty level based on the student's learning ability, the higher the learning ability, the higher the difficulty level determined. This function then calculates each student's unmastered knowledge items by the correctness of the results acquired in intelligent correcting.
3. *Generation of personalized assessment tasks*: First the function needs to calculate the number of recommended questions, including two aspects: the number of questions for each type and the number of questions for each unmastered knowledge item. Then, the function generates personalized assessment tasks, which are matched to students'

learning abilities, based on the appropriate difficulty, unmastered knowledge items, and the number of recommended questions.

4. *Forgetfulness-based assessment tasks recommendations*: In this function, forgetting functions are used to predict students' memory retention. Over time, when the retention of knowledge items is less than a threshold, the system determines that the student has forgotten that part of the knowledge items, and the function provides assessment tasks related to the forgotten knowledge items.

4 System Design of Intelligent Personalized Assessment Tasks Recommendation System

Based on the students' learning abilities, the intelligent personalized assessment tasks recommendation system selects appropriate learning resources to compensate for the students' deficiencies of knowledge, can better assist teachers in teaching, and also increase students' confidence and interest in learning.

4.1 Data Processing Component

In this component, three functions are included: question bank preparation, intelligent auto-generating test paper, and intelligent correcting. The function bank preparation is used to generate the question bank and analyze questions. The function of intelligent auto-generating test paper is used to automatically generate assignments according to teaching requirements. The function of intelligent correcting is used to correct the answers obtained from students.

Question Bank Preparation

This function consists of two parts: question bank generation and question analysis. The question bank generation is used to collect the questions through text, image, or voice. The question analysis is used to identify knowledge items and the difficulty of questions. More detailed information on the above parts is shown below.

1. *Question bank generation*: This function supports text, image, and voice input to questions and answers, which are converted to text and automatically identifies the type of problem such as fill-in-the-blank, application problems. This step contains the technology of speech recognition and optical character recognition. Speech recognition can be implemented by vector quantization [12], SVM [13], etc. Optical character recognition can be implemented by convolutional neural networks [14] ResNet,VGGNet [15], etc. The function provides two modification options: automatic system modification completing missing text based on semantics, e.g., RNN [16], and manual modification to ensure the accuracy of the data after system modification. Finally, the modified data is stored in the question bank.
2. *Question analysis*: This function processes the text data to automatically obtain the knowledge items and difficulty of the questions. In tradition, due to the complexity of the questions, the knowledge items and difficulty of questions are mainly labeled by manual approaches. In this function, through keyword extraction technology,

e.g., TextRank [17], the keywords in question are obtained. Then, this function automatically identifies the knowledge items in the question based on the keywords. According to the knowledge items, the difficulty of the question is calculated. The process of question bank preparation is shown in Fig. 2.

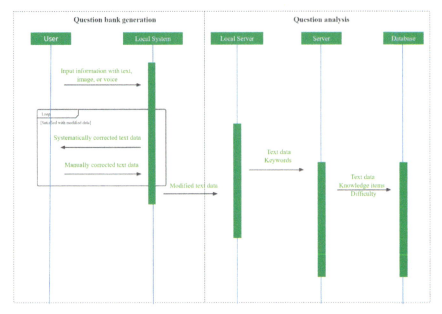

Fig. 2. The process of question bank preparation

Question bank supports multidimensional input and automatically identifies the knowledge items and labels the difficulty of the questions, so that the question bank can be automatically expanded at a low cost. Then, based on the information in the question bank, an initial assessment task is generated.

Intelligent Auto-generating Test Paper

The intelligent auto-generating test paper function uses the grouping algorithm, e.g., genetic algorithm [20], to select the questions to form a single assessment task. The function captures two requirements: the expected difficulty of assessment tasks and coverage of knowledge items. The difficulty refers to the integrated difficulty of the assessment tasks, e.g., the average value of the difficulty of questions. The function generates assignments in which the difficulty of the questions meets a positive-term distribution so that the student's learning ability is fully examined. Based on the teaching plan, the function selects the knowledge items that need to be mastered, and attaches a priority to each knowledge item, the higher the priority, the more number of questions containing this knowledge item in initial assessment tasks. This original assessment task is used as an important basis for examining the student's learning ability.

Intelligent Correcting

Based on the technology of handwriting recognition, this intelligent correction function evaluates the correctness of answers. Handwriting recognition refers to the technique of recognizing handwritten characters, e.g., optical character recognition from an image and converting them into symbols that can be recognized by the system. According to the symbols obtained, this function then evaluates the answers to subjective and objective questions. Objective questions' answers contain only results, such as fill-in-the-blank questions, so intelligent correcting function only needs to judge the correctness of the results. Subjective questions' answers require the student to provide not only the results but also the detailed steps to solve the problem, such as an application question, so intelligent correcting function needs to evaluate the correctness of both the steps and results from students' answers.

The intelligent correcting function allows us to obtain the correctness of the results of each objective question and the correctness of the steps and results of each subjective question. Then, according to the situation of the assessment tasks, the function provides personalized assessment tasks recommendations for each student.

4.2 Personalized Assessment Tasks Recommendations Components

In this subsection, we describe the process of recommending assignments based on the student's learning ability to learn, which is divided into the following four steps: evaluation of students' learning abilities, calculation of appropriate difficulty, and mastery of knowledge items.

Evaluation of Students' Learning Abilities

The difficulty of each question is calculated by the question bank preparation function, and through the intelligent correcting function, the correctness of students' answers is acquired. Based on the difficulty of the questions and the correctness of the answers, students' learning abilities are evaluated.

To evaluate students' learning abilities, personal assessment tasks recommendations components choose item response theory, also known as latent trait theory, to model students' learning abilities. Objective questions only need to determine the correctness of the results, while subjective questions need to determine the correctness of the results and steps, hence based on their different characteristics, different models are chosen to evaluate students' learning abilities.

The process of evaluating the student's learning ability is shown in Fig. 3, where we first obtain the correctness of the student's answers and the difficulty of each question. Then, this function determines the type of each question. This function then selects the student's learning ability evaluation model based on the question type. Finally, based on the difficulty of the questions, the correctness of the answers, and the student's learning ability evaluation model, this function evaluates the student's learning ability on the subjective and objective questions separately.

Calculation of Appropriate Difficulty and Mastery of Knowledge Items

This function calculates the appropriate difficulty and mastery of each knowledge item for each student, which serves as the basis for the recommendation of personalized assessment tasks. The process is shown in Fig. 4. The first is the calculation of the appropriate difficulty level matching students' learning abilities. Previously, the student's learning ability for each type of question is calculated, and through building a mapping model of assessment tasks' difficulty and student's learning ability, the appropriate assessment tasks' difficulty is obtained.

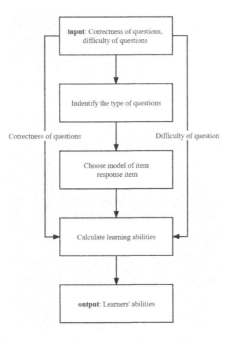

Fig. 3. Flow chart of evaluation of student's learning ability

The next step is to calculate the mastery of knowledge items. Usually, the error rates of knowledge items are an important indicator of the mastery of knowledge items, so this function calculates the error rate of each type of question and each knowledge item. For different types of questions, the calculation criteria is different. For objective questions, because they only have results, when the question is incorrect, this function determines that all knowledge items corresponding to the question have not been mastered. For objective questions, this function determines the mastery of each knowledge item based on the solving steps, and if the step is wrong, the knowledge item corresponding to that step is determined as not mastered. Finally, this function calculates the error rate for each knowledge item and for questions of each type.

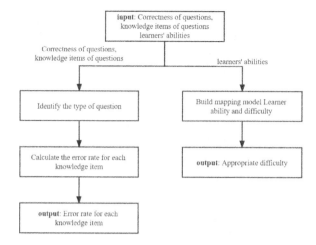

Fig. 4. Flow chart for calculating the appropriate difficulty and mastery of knowledge items

Generation of Personalized Assessment Tasks

To generate a personalized recommendation for an assessment task, we need to rationalize the number of questions in the assessment task. Firstly, the number of objective questions or subjective questions in the next recommendation is shown in Eq. (1).

$$N_i = \frac{m \times E_i}{E_o + E_s} \quad i = o \text{ or } s \tag{1}$$

In Eq. (1), m means a total number of questions in next the assignment. E_i means the error rate of objective or subjective questions. N_i means the number of objective or subjective questions in the next recommendation. The number of each unmastered knowledge item of objective questions contains in the assignment for the next recommendation, as shown in Eq. (2).

$$N_{k,j} = \frac{N_k \times e_{o,j}}{\sum_{i=1}^{n} e_{o,i}} \quad k = o \text{ or } s \tag{2}$$

In Eq. (2), $e_{o,j}$ means the error rate of knowledge item j, N_{kj} means the number of knowledge items of objective questions or subjective questions contains in the assignment for the next recommendation.

Then, we provide each student with assessment tasks appropriate to his or her learning ability, with the following constraints, reinforcement learning [18], collaborative filtering [19], etc.

1. This function generates assessment tasks for each student at a level of difficulty equivalent to the appropriate difficulty matching their learning ability.
2. This function generates for each student the number of questions contained in the assessment task equal to the number obtained in Eq. (1).

3. The number of knowledge items contained in each type of question in the assessment task for each student is equal to the number calculated by Eq. (2).

In this subsection, this function calculates the number of questions recommended for each type of question, the number of questions recommended for each knowledge item and also combines the appropriate difficulty level to generate homework recommendations for each student. Then this function recommends assessment tasks to help students consolidate their knowledge items related to forgetting.

Forgetfulness-based Assessment Tasks Recommendations

Forgetfulness refers to the inability or error of remembered knowledge items. Because students have different physiological characteristics and life experiences, they may have different memory characteristics, i.e., the forgetting process is different for each individual. We choose forgetting functions to fit the forgetting process of each student, as shown in Eq. (3).

$$\omega = \frac{a}{(bt+1)^c} \qquad (3)$$

In Eq. (3), ω shows the percentage of personal memory retention, t represents the time since learning the knowledge, a indicates the degree of original learning, b means a scaling constant, and c manifests the rate of forgetting. The recommendation process for forgotten knowledge items is shown in Fig. 5.

This function divides the knowledge by day, and the knowledge learned on that day is grouped into one module. Over time, when a student's knowledge items in the module are judged to be forgotten, this function recommends assessment tasks based on forgotten knowledge items.

After the forgetting function simulation, if the student's memory level for the knowledge item is lower than the set threshold, this function recommends this module of knowledge items to this student. The recommendations are divided into two steps. The first step is to recommend questions based on students' unmastered knowledge items, and the second step is to recommend questions based on all knowledge items in this module to prevent forgetting the mastered knowledge items.

Recommendation with unmastered knowledge items. This function stores the unmastered knowledge items that when studying this module and the difficulty of the assessment tasks appropriate to the student's learning ability. When the knowledge items of this module are judged to be forgotten, Eqs. (1) and (2) is used to calculate the number of questions that need to be recommended for each type of question and each knowledge item. Based on the appropriate level of difficulty, and the number of questions, this function recommends the assessment tasks.

To prevent students from forgetting about mastered knowledge items, this function also recommends questions related to mastering. If the number of times a knowledge item is judged to be mastered exceeds the threshold, this function considers the student to have fully mastered the knowledge item and does not recommend it, removing it from the mastered knowledge set. This function then recommends assignments for students based on the most appropriate level of difficulty and mastery.

Based on the student's performance, this function calculates the student's learning ability and unmastered knowledge as data for the next forgetting recommendation.

4.3 System Analysis

Firstly, the framework design of the intelligent assessment tasks recommendation system is a hybrid recommendation. For knowledge items that are learned recently, this framework calculates the student's learning ability based on the correctness of assignments and recommends assignments based on mastery of knowledge items and suitable difficulty.

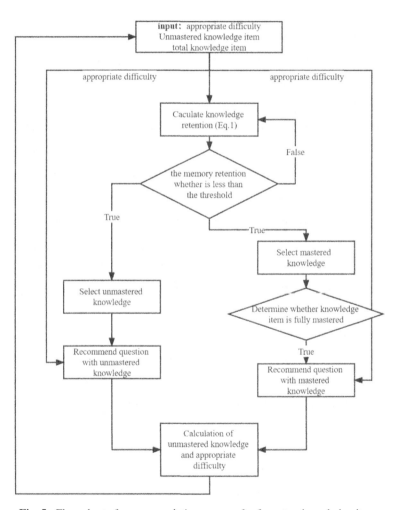

Fig. 5. Flow chart of recommendation process for forgotten knowledge items

For forgotten knowledge items that are learned a long time before, this framework recommends assignments based on the student's mastery of this part of the knowledge items and the appropriate level of difficulty in the past. Meanwhile, this framework dynamically estimates the learner's learning ability based on the learner's recent data. Secondly, the framework design of the intelligent assessment tasks recommendation system helps to free the teacher from the heavy workloads by incorporating computer technology, which is capable of extracting the knowledge items, calculating the difficulty of questions, automatically designing assessment tasks, and automatically correcting assignments. Thirdly, the framework design of the intelligent assessment tasks recommendation system also requires recent data in the process of recommending assignments for students and allows for dynamic adjustment of the recommendation strategy. Finally, based on their learning ability and forgetfulness, this system can recommend an appropriate level of difficulty with unmastered knowledge items for the student.

5 Conclusion

The intelligent assessment tasks recommendation framework design for personalized learning proposed in this paper incorporates current mainstream modern educational technologies, such as knowledge item recognition, intelligent auto-generating of test papers, and intelligent auto-correcting on assessment tasks. Through considering multiple aspects, i.e., students' learning ability, students' knowledge mastery, the difficulty of assessment tasks, and students' forgetting characteristics, the proposed framework in this paper provides a solution to improving lecturers' teaching quality and students' learning performance in personalized learning.

Acknowledgments. We would like to thank the editors and reviewers for their constructive comments and suggestions to enhance the quality of the paper. This work has been supported in part by the National Natural Science Foundation of China under Grant No. 62006090, and the Fundamental Research Funds for the Central Universities, CCNU under Grant No.3110120001.

References

1. Nabizadeh, A.H., Leal, J.P., Rafsanjani, H.N., Shah, R.R.: Learning path personalization and recommendation methods: a survey of the state-of-the-art. Expert Syst. Appl. **159**, 113596 (2020)
2. Hussain, M., Zhu, W., Zhang, W., Abidi, S.M.R., Ali, S.: Using machine learning to predict student difficulties from learning session data. Artif. Intell. Rev. **52**(1), 381–407 (2018). https://doi.org/10.1007/s10462-018-9620-8
3. Benhamdi, S., Babouri, A., Chiky, R.: Personalized recommender system for e-Learning environment. Educ. Inf. Technol. **22**(4), 1455–1477 (2016). https://doi.org/10.1007/s10639-016-9504-y
4. Karga, S., Satratzemi, M.: A hybrid recommender system integrated into LAMS for learning designers. Educ. Inf. Technol. **23**(3), 1297–1329 (2017). https://doi.org/10.1007/s10639-017-9668-0

5. Khanal, S.S., Prasad, P.W.C., Alsadoon, A., Maag, A.: A systematic review: machine learning based recommendation systems for e-learning. Educ. Inf. Technol. **25**(4), 2635–2664 (2019). https://doi.org/10.1007/s10639-019-10063-9
6. Chen, Y.-H., Tseng, C.-H., Huang, C.-L., Deng, L.Y., Lee, W.-C.: Recommendation system based on rule-space model of two-phase blue-red tree and optimized learning path with multimedia learning and cognitive assessment evaluation. Multimed. Tools Appl. **76**(18), 18237–18264 (2016). https://doi.org/10.1007/s11042-016-3717-3
7. Kolekar, S.V., Pai, R.M., Manohara Pai, M.M.: Rule based adaptive user interface for adaptive E-learning system. Educ. Inf. Technol. **24**(1), 613–641 (2019)
8. Shu, J., Shen, X., Liu, H., Yi, B., Zhang, Z.: A content-based recommendation algorithm for learning resources. Multimedia Syst. **24**(2), 163–173 (2017). https://doi.org/10.1007/s00530-017-0539-8
9. Zhou, Y., Huang, C., Hu, Q., Zhu, J., Tang, Y.: Personalized learning full-path recommendation model based on LSTM neural networks. Inf. Sci. **444**, 135–152 (2018)
10. Yera, R., Martínez, L.: A recommendation approach for programming online judges supported by data preprocessing techniques. Appl. Intell. **47**(2), 277–290 (2016). https://doi.org/10.1007/s10489-016-0892-x
11. Dwivedi, P., Kant, V., Bharadwaj, K.K.: Learning path recommendation based on modified variable length genetic algorithm. Educ. Inf. Technol. **23**(2), 819–836 (2017). https://doi.org/10.1007/s10639-017-9637-7
12. Hu, H.T., Chou, H.H., Lee, T.T.: Robust blind speech watermarking via fft-based perceptual vector norm modulation with frame self-synchronization. IEEE Access **9**, 9916–9925 (2021)
13. Tamiru, N.K., Tekeba, M., Salau, A.O.: Recognition of Amharic sign language with Amharic alphabet signs using ANN and SVM. Vis. Comput. **38**(5), 1703–1718 (2021). https://doi.org/10.1007/s00371-021-02099-1
14. Xu, X., Tan, M., Corcoran, B., et al.: 11 TOPS photonic convolutional accelerator for optical neural networks. Nature **589**(7840), 44–51 (2021)
15. Wang, J., Sun, K., Cheng, T., et al.: Deep high-resolution representation learning for visual recognition. IEEE Trans. Pattern Anal. Mach. Intell. **43**(10), 3349–3364 (2020)
16. Yang, S., Gong, Z., Ye, K., Wei, Y., Huang, Z., Huang, Z.: EdgeRNN: a compact speech recognition network with spatio-temporal features for edge computing. IEEE Access **8**, 81468–81478 (2020)
17. Tao, Y., Cui, Z., Jiazhe, Z.: Research on keyword extraction algorithm using PMI and TextRank. In: 2019 IEEE 2nd International Conference on Information and Computer Technologies (ICICT), pp. 5–9. IEEE, Kahului (2019)
18. Sakurai, K., Togo, R., Ogawa, T., Haseyama, M.: Listener recommendation for artist based on knowledge graph and reinforcement learning. In: 2021 IEEE 10th Global Conference on Consumer Electronics (GCCE), pp. 202–203. IEEE, Kyoto (2021)
19. Wu, L.: Collaborative filtering recommendation algorithm for MOOC resources based on deep learning. Complexity **2021**, 5555226 (2021)
20. Wang, J., Shen, L.: Research and design of intelligent test paper system based on genetic algorithm. In: 2020 International Conference on Big Data & Artificial Intelligence & Software Engineering (ICBASE), pp. 292–295. IEEE, Bangkok (2020)

Assessing Graduate Academic Scholarship Applications with a Rule-Based Cloud System

Yongbin Zhang[1], Ronghua Liang[1], Yuansheng Qi[1], Xiuli Fu[2(✉)], and Yanying Zheng[3]

[1] Beijing Institute of Graphic Communication, Beijing, China
{zhangyongbin,liangronghua,yuansheng-qi}@bigc.edu.cn
[2] Beijing Institute of Petrochemical Technology, Beijing, China
fuxiuli@bipt.edu.cn
[3] Beijing University of Agriculture, Beijing, China
huaxue@bua.edu.cn

Abstract. Academic scholarships motivate graduates to work hard. However, the tedious application process frustrates students and brings adverse effects. The paper presents a rule-based cloud computing system to assess academic scholarship applications. The criteria can be encoded with if-then expressions. The rule-based system enables the administrator to add new standards or update existing ones flexibly. Students can upload their achievements through a web browser on different devices. Graduates know the mark of each item input into the system. The system calculates final grades automatically. Members of evaluation committees can review and return applications to applicants. All criteria are available for students. The transparency encourages students to participate in applying for scholarships. The experimental results show that the designed system was helpful in evaluating scholarship applications. Students spent less time applying for academic scholarships with the rule-based cloud system than with traditional approaches.

Keywords: Assessing · Academic scholarship · Cloud computing · Rule-based system

1 Introduction

Academic scholarships for graduates encourage and motivate students to achieve outstanding academic performance. In contrast to financial aid, which supports low-income families, academic scholarships gauge academic achievements. Students have to provide supportive documents during their application processes. However, most application processes are complex and tedious, which frustrate applicants.

Higher education should provide students with non-complex administrative processes to serve their non-academic needs. Student motivation is affected when faced with the frustration of tedious, complex, and laborious procedures [1]. For example, students can become demotivated and frustrated when having difficulties with the registration of modules, timetable changes, assessment deferrals, or fees negotiations.

Therefore, this paper presents a rule-based cloud system evaluating academic scholarship applications for graduates. The system simplifies the application process and motivates graduates to participate in educational scholarship applications.

2 Literature Review

2.1 Graduate Scholarship

With the increasing demand for knowledgeable and skillful employees, more and more students pursue higher degrees or even doctoral degrees. There is a skills gap in the labor force, and employers have complained that they cannot find the graduates needed [2]. One key factor for employers is looking for graduates with higher degrees when recruiting employees. The requirement for higher degrees incents students to continue their education. The expansion in higher education put stress on universities. Then increasing tuition fees is an effective way to overcome the difficulties that universities face. The rising tuition cost may keep some students from achieving their degrees.

To help graduate students attain their goals, enterprises and governments provide various types of financial assistance for graduate students. Financial aid is particularly critical for graduates from low-income families. State-funded grants have been an increasingly important source of financial help [3]. However, with the development of the global economy and increase in income, merit-based scholarship has been increasing faster the traditional need-based financial aid. The need-based scholarship aims to provide financial assistance for students from low-income families. The merit-based scholarship rewards students according to their academic performance.

In China, all students have to pay tuition fees for their higher education since 1998 [4]. All universities have charged graduates on the full scale [5]. The graduate education system is to cultivate a talented workforce for Chinese economic development. The Chinese government emphasizes quality and equality in its graduate education system. Therefore, our government has provided more financial aid for graduate education, including low-income help, excellent academic performance awards, outstanding outcomes, excellent leader awards, and innovative plans.

In this paper, we focus on the evaluation of academic scholarship applications. One reason is that more graduates apply for academic scholarships than for low-income financial aid. Another reason is that more criteria need for academic scholarships than other financial aids. If our presented rule-based system works to evaluate academic scholarships, we can use the system to assess other financial aids. Also, academic scholarships require the applicants to show their achievements publicly. Therefore, it is appropriate to adopt the cloud system so that all students and staff can view the information.

2.2 Motivation for Academic Scholarship

The national or state-funded academic scholarship aims to reward graduate students who achieved outstanding performance. Academic records determine the scholarship instead of financial needs [6]. The money works as a goal to motivate students. To obtain the aim, students have to present their higher academic achievement. Therefore, academic scholarships encourage graduate students to work hard for excellent performance.

Some students may view academic scholarship as intrinsic for the enjoyment of educational activities. Other students may be extrinsically motivated to obtain some reward. However, complex administrative processes may demotivate and frustrate students [6]. For example, if the application processes are difficult or tedious, students may feel frustrated. Higher education should provide students with a convenient and straightforward administrative process for non-academic needs [1]. Therefore, this paper provides an effective and efficient way to evaluate graduate academic scholarship applications.

2.3 Rule-Based System

A Rule-based system belongs to expert systems. One crucial feature of the rule-based system is that specialist knowledge can be encoded with rules. A rule consists of an antecedent and a consequent. A criterion can be written in a production form as follows:

$$C \rightarrow A$$

Moreover, the above rule can be expressed in a human-friendly way as follows:

$$IF\ C,\ THEN\ A$$

The condition can be a single condition like a simple equal judgment, IF a = 'A'. Of course, logical operators with simple requirements compose a complex one.

$$IF\ a = \text{`A'}\ and\ b \geq 3$$

And the action part can comprise a single action or more actions. For example, we can assign a specified value to the variable.

$$IF\ a = \text{`A'}\ and\ b \geq 3\ THEN\ point = 10$$

A rule-based system consists of a set of human-coded rules. Although the most straightforward artificial intelligence system, rule-based systems are one of the best approaches for encoding human expertise [7].

The benefits of a rule-based system include simplicity, efficiency, and explicit expression. There are many differences between rule-based AI and machine learning systems. Firstly, rule-based AI models are deterministic. And machine learning systems are probabilistic. Secondly, a rule-based system does not require a large amount of data to train the model as a machine learning system does.

Concrete and determining criteria are needed to ensure the same results for the same input when undergraduates apply for scholarships. Therefore, a rule-based system is appropriate for assessing scholarship applications.

2.4 Cloud Computing

Cloud computing has attracted attention from all fields because of the potential benefits over traditional computing. For example, cloud computing models provide three to five times the cost advantage for business applications and even more for consumer applications. With the economic slowdown and unprecedented budget cuts, universities seek cloud computing services as one of the cutting-cost measures [8].

Cloud computing is a model for enabling ubiquitous, convenient, on-demand network access to a shared pool of configurable computing resources that can be rapidly provisioned and released with minimal management effort or service provider interaction [9]. Although a uniform definition of cloud computing does not exist, different meaning share the typical components of clouds, including on-demand self-service, broad network access, resource pooling, rapid elasticity, and measured service. Users can consume cloud computing services through three models, including Software as a Service (SaaS), Platform as a Service (PaaS), and Infrastructure as a Service (IaaS) [10]. Different service models provide different levels of abstraction. For example, users can assess an application over the Internet from various devices through a web browser with the SaaS model. Developers can develop applications on a cloud platform with the PaaS model. The IaaS model offers clients virtual servers. The SaaS is the most popular service model in higher education [8].

There are types of deployment models for cloud computing. If cloud resources are exclusively offered for a specific organization, it is called a private cloud. On the other hand, if the cloud resources are available to the public, it is a public cloud. Different deployment models are derived from the private and public clouds. The community cloud stands between the public and the private clouds. A community cloud works like a private cloud, but resources are exclusive to two or more organizations rather than a single one. And the hybrid cloud composes at least two cloud models above [11].

The critical characteristics of cloud computing include on-demand self-service, broad network access, resource pooling, rapid elasticity, and measured service. These features satisfy our requirements for assessing graduate scholarships. Students need to input data into the system conveniently. Cloud system improves the transparency of the application information. On the other hand, universities should protect the confidence of undergraduate students.

We adopted the private cloud deployment model and the software as a service model. With a private cloud deployment model, the system is only accessible to undergraduate students and staff at our university. And all information is stored in our data center. Both features sustain the security demand. With the SaaS service model, users can utilize the system on different devices anytime. The SaaS model provides max flexibility for all users.

3 The Architecture of the Rule-Based Cloud System

This scholarship evaluation system comprises three parts, the frontend, the backend, and the cloud servers. The cloud service provider offers the hardware and software platforms, such as database servers, application servers, rule engine servers, and web servers. With cloud service, we can focus on the functions and processes of the evaluation system rather than the infrastructure and development platforms.

The backend includes management functions of the system, such as scholarship categories management, premises, and rules management. For example, a system administrator can add a new academic scholarship into the scholarship category. The existing type of scholarships could be updated or deleted. The rule management model satisfies the requirements of different scholarship evaluations and provides flexibility for schools. New criterion could be added to the system. And existing rules could be updated without affecting the application processes.

The front end provides all functions for students and members of different levels of validation. Graduates can apply for academic scholarships through a web browser from anywhere. Students input each item of their achievement. For each item, students set its properties by selecting from the premises database without guessing or keyboard input. This selecting feature reduces the probability of error inputs.

The architecture of the rule-based could system is shown in Fig. 1.

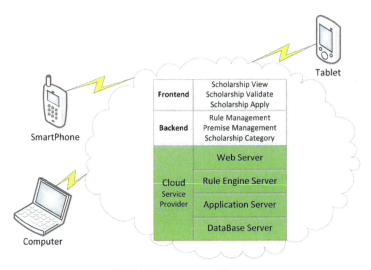

Fig. 1. The system architecture.

The administrators should set scholarship categories, premises, and rules before graduates apply for academic scholarships. All academic scholarships, antecedents, and criteria would be stored in the rule database.

Graduates could apply for different academic scholarships published by the administrators. When students input an academic achievement into the system, students select and set the properties for each item. All input from students will go to the application database. The fork of the rule database and the application database enhances the reuse of the rules saved. The separation also optimizes operations for rules management and application management.

After graduates submit their applications, members of scholarship commitment can check the applications. If the members have any doubt, they can return the application to the applicants. Then the applicants could update and resubmit their applications.

All students and staff can review all scholarship applications to improve transparency. The scholarship application review model has three objectives. One aim is to ensure every graduate has the right to apply for scholarships. No single applicant will be ignored because every submission is published online. Another goal is to keep equality among

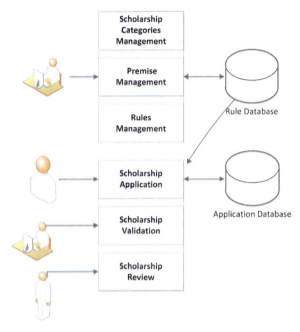

Fig. 2. The system functions.

all applicants. The same criteria are applied to every applicant equally. The third purpose is to motivate every graduate to work hard and attain better performance. The different roles and functions are shown in Fig. 2.

4 Experiment

There are different academic scholarships for graduates in Chinese universities. Types of academic scholarships vary among national universities and local universities. The criteria for the same scholarship in a university may be different from one department to another. A department could modify the requirements every year according to the situation. However, the standards can be coded with rules. For example, the criteria for the national academic scholarship for the graduate enrolled in 2019 consisted of three parts at the Beijing Institute of Graphic Communication (BIGC). The first component included published academic papers; the second part involved granted patents; the third category comprised educational prizes. Each item was assigned a mark. For example, for a published academic paper, the function was available.

$$f(y,x) = \begin{cases} 50, y \in [\textit{the first author}] \wedge (x \in [\textit{Ei journal}] \cup x \in [\textit{Sci journal}]) \\ 20, y \in [\textit{the second author}] \wedge (x \in [\textit{Ei journal}] \cup x \in [\textit{Sci journal}]) \\ 20, y \in [\textit{the first author}] \wedge (x \in [\textit{Chinese indexed journal}]) \\ 10, y \in [\textit{the second author}] \wedge (x \in [\textit{Chinese indexed journal}]) \\ 10, y \in [\textit{the first author}] \wedge (x \in [\textit{non indexed journal}]) \\ 5, y \in [\textit{the second author}] \wedge (x \in [\textit{non indexed journal}]) \end{cases} \quad (1)$$

To encode the above function with IF-THEN rules, we first transferred the condition into atom ones.

$A_1 \rightarrow$ the first author.
$A_2 \rightarrow$ the second author.
$P_1 \rightarrow$ Sci journal.
$P_2 \rightarrow$ Ei journal.
$P_3 \rightarrow$ Chinese indexed journal.
$P_4 \rightarrow$ non indexed journal.

Then we designed the following rules:

IF $A_1 \wedge P_1$ THEN point = 50.
IF $A_1 \wedge P_2$ THEN point = 50.
IF $A_2 \wedge P_1$ THEN point = 20.
IF $A_2 \wedge P_2$ THEN point = 20.
IF $A_1 \wedge P_3$ THEN point = 20.
IF $A_2 \wedge P_3$ THEN point = 10.
IF $A_1 \wedge P_4$ THEN point = 10.
IF $A_2 \wedge P_4$ THEN point = 5.

When graduates apply for an academic scholarship, they input valid achievements from three categories. The system would determine the points for each item input. And the final grade of an applicant was calculated by adding points from all items with the following formulate.

$$f(x) = \sum_{i=1}^{3} \sum_{j=1}^{n} point_{ij}, i \in [1, 3], j \in [1 \ldots n] \quad (2)$$

We adopted this system to evaluate graduate academic scholarship applications in 2021. Graduates from grade-2018, 2019, and 2020 participated in the process.

5 Results

All graduates from three grades applied for academic scholarships for the 2021 year. All scholarships demanded that all outcomes or achievements should be gained or produced during 2021. There were 21 students from grade 2018, 40 from 2019, and 41 from 2020. All students participated in the excellent performance scholarship. The superb performance scholarship had three classes. The first class had twenty percent of the total students, thirty percent for the second class, and fifty percent for the third class.

Three graduates applied for the national academic scholarship, one from each grade, because this type of scholarship had a limit rate for each group.

Twenty students applied for the outstanding academic achievement scholarship. This type of scholarship had no limit for student numbers.

Twelve students applied for the innovative, practical scholarship. This kind of scholarship encouraged graduates to design creative plans for industries with a five percent of all students enrolled limit set.

Two graduates applied for the excellent competition reward scholarship. This type of scholarship motivated students to participate in all levels of academic completion events. And there was no limit set for this kind of scholarship.

All applicants input and upload their information into the cloud system. For each item input, the system could show them the corresponding points of the item according to the rules. The system calculated and displayed the summative grade dynamically. Graduates submitted all applications within three days. It usually took students one week to submit their material without the system.

After an applicant applied, the scholarship committee of students would verify the validation of each item. If there were any doubt or error, the student committee would disapprove of the application, and the applicant could modify and resubmit the application. If the committee approved the application, the department committee, including administrative staff, teachers, and selected students, would review and discuss the application. After the department commitment made the decisions, the university commitment would go over all the material and form the final result. All work was done on the cloud system.

6 Discussion

The presented rule-based cloud system allows graduates to submit their application materials online. Applicants could view the point for each item input and the final summative points of their application. The system facilitated graduates in academic scholarship applications. The characteristics included easy usage, transparent rules, and instant calculation. Experimental results showed graduates finished their submissions in a shorter time with the system than with traditional ways. The rule-based system allowed students to focus on their achievement according to the criteria. Therefore the system could motivate students to participate in the academic scholarship application.

Although the system shorted the time needed for applications for different academic scholarships, students had to collect material related to their achievements last year. It may be tedious to look for what happened in the past. Therefore, more work should be done to alleviate students' workload.

7 Conclusion

The rule-based cloud system evaluated academic scholarships applications for graduate students. The system enabled students to input their documents online and to view the grade for each item input. Final summative points were calculated automatically according to the standards. The system lessened the burden on students during their applications. The rule-based cloud system could motivate students to participate in academic scholarship applications by providing transparent public rules and easy usage. We will improve the information collection process to reduce the workload in the future.

Acknowledgment. Yongbin Zhang thanks the Beijing Municipal Education Commission and the Ministry of Education of the People's Republic of China (202102122003) for supporting this research.

References

1. Nukpe, P.: Motivation: theory and use in higher education. Invest. Univ. Teach. Learn. **8**, 11–17 (2012)
2. Hesketh, A.J.: Recruiting an elite? Employers' perceptions of graduate education and training. J. Educ. Work. **13**(3), 245–271 (2000)
3. Heller, D.E., Marin, P.: Who should we help? The negative social consequences of merit scholarships (2002)
4. Marcucci, P.N., Johnstone, D.B.: Tuition fee policies in a comparative perspective: theoretical and political rationales. J. High. Educ. Policy Manag. **29**(1), 25–40 (2007)
5. Wei, W., Xima, Y.: Research on accunting development cost per graduate student in university. Can. Soc. Sci. **13**(1), 11–15 (2017)
6. Scott-Clayton, J.: On money and motivation a quasi-experimental analysis of financial incentives for college achievement. J. Hum. Resour. **46**(3), 614–646 (2011)
7. Hayes-Roth, F.: Rule-based systems. Commun. ACM **28**(9), 921–932 (1985)
8. Britto, M.: Cloud computing in higher education. Libr Student J. **7** (2012)
9. Mell, P., Grance, T.: The NIST definition of cloud computing (2011)
10. Rani, D., Ranjan, R.K.: A comparative study of SaaS, PaaS and IaaS in cloud computing. Int. J. Adv. Res. Comput. Sci. Softw Eng **4**(6) (2014)
11. Goyal, S.: Public vs private vs hybrid vs community-cloud computing: a critical review. Int. J. Comput. Netw. Inf. Secur. **6**(3), 20–29 (2014)

AI-Based Visualization of Voice Characteristics in Lecture Videos' Captions

Tim Schlippe[1(✉)], Katrin Fritsche[2], Ying Sun[2], and Matthias Wölfel[3]

[1] IU International University of Applied Sciences, Erfurt, Germany
tim.schlippe@iu.org
[2] University Jena, Jena, Germany
[3] Karlsruhe University of Applied Sciences, Karlsruhe, Germany

Abstract. More and more educational institutions are making lecture videos available online. Since 100+ empirical studies document that captioning a video improves comprehension of, attention to, and memory for the video [1], it makes sense to provide those lecture videos with captions. However, studies also show that the words themselves contribute only 7% and how we say those words with our tone, intonation, and verbal pace contributes 38% to making messages clear in human communication [2]. Consequently, in this paper, we address the question of whether an AI-based visualization of voice characteristics in captions helps students further improve the watching and learning experience in lecture videos. For the AI-based visualization of the speaker's voice characteristics in the captions we use the *WaveFont* technology [3–5], which processes the voice signal and intuitively displays loudness, speed and pauses in the subtitle font. In our survey of 48 students, it could be shown that in all surveyed categories—*visualization of voice characteristics*, *understanding the content*, *following the content*, *linguistic understanding*, and *identifying important words*—always a significant majority of the participants prefers the *WaveFont* captions to watch lecture videos.

Keywords: Closed captions · Subtitles · Speech processing · Natural language processing · Digital humanities · AI in education

1 Introduction

Access to education is one of people's most important assets, and ensuring inclusive and equitable quality education is goal 4 of United Nations' Sustainable Development Goals [6]. Distance learning, in particular, can create education in areas where there are no educational institutions or in times of a pandemic. There are more and more distance learning opportunities worldwide and challenges like the physical absence of the teacher and the classmates or the lack of motivation of the students are addressed with technical solutions like video conferencing systems [7] and gamification of learning [8]. The research area "AI in Education" addresses the application and evaluation of artificial intelligence (AI) methods in the context of education and training [9]. For instance, it deals with sentiment analysis to classify students' comments [10], natural language processing based tutoring systems [11], automatic short answer grading [12, 13], recommender systems [14] or conversational AI systems, which optimally prepare students for their exams [15, 16]. To overcome the linear structure of presentations and video lectures, [17] proposes to automatically generate a dialog system from slide-based presentations which can dynamically adapt to student requests. The capacity of the system is still limited, but its usefulness has already been confirmed by learners and lecturers alike.

Even though AI-based approaches show considerable potential, video lectures—along with classroom and online presentations—remain one of the dominant methods for conveying information. Since many educational institutions—public and private—already provide lecture videos online, even small advances in improving the effectiveness of video lectures will have a big impact on learners' knowledge acquisition. As understanding the spoken word is not always optimal due to a noisy recording (lecturers are not professional sound engineers), noisy environment of the viewer (e.g., on a train), or due to temporary or permanent disabilities, subtitling and captioning are used[1]. In addition, more than 100 empirical studies document that captioning a video improves comprehension of, attention to, and memory for the video [1]. However, conventional captions and subtitles have not evolved for decades. They still reflect *what* is spoken—not *how* it is spoken—i.e., no information about loudness, intonation, pauses, lengths, and emotions. We believe that potential is not being exploited here, as studies show that the words themselves contribute only 7% and how we say those words with our tone, intonation, and verbal pace contributes 38% to making messages clear in human communication [2]. Consequently, in this paper, we address the question of whether an AI-based visualization of voice characteristics in captions helps students further improve the watching and learning experience in lecture videos.

In the next section, we will present the latest approaches of other researchers to captioning and subtitling, as well as to the visualization of non-textual information, such as an utterance's emotion or prosody. In Sect. 3 we will introduce our technology *WaveFont* which we use to map acoustic features to font characteristics. Section 4 will describe the experimental setup for our user study. The study and the results are outlined in Sect. 5. We will conclude our work and suggest further steps in Sect. 6.

2 Related Work

Several studies show that educational videos should meet some requirements—be of shorter length, contain high quality image and text components, etc. [18, 19]. The effect of captions and subtitles in those videos is described as supporting the understanding of thematic content as well as improving literacy and language skills [20–23]. [24] and [25] report that particularly in the university context students prefer videos with captions.

Studies which focus on captioning and subtitling in general are concerned with their placement and design [26, 27]. An eye-tracking study indicates that these parameters can affect reading time and the visual perception of the image [28]. Traditional captions and subtitles are limited to telling the audience *what* is merely being said instead of

---

[1] In this research we refer to interlingual translation as *subtitles* and transcription in the same language as *captions*.

how it is being said [29]. These methods do not present information beyond verbatim dialogue such as emotional expressions [30] and can lead to communication problems for the receiving audience [31]. Consequently, some studies assert the benefits of captions for the viewers, to make the material more understandable to them [32]. [3] state that creative captions and subtitles can benefit a wide range of people, not only deaf and hard-of-hearing. [33] investigates the use of creative subtitles through emojis and emoticons as well as reports that its use furthers the function of standard subtitling, allowing for tone of voice and emotions to be conveyed to the target audience.

To improve comprehension and to include non-textual information, such as emotion or prosody in an utterance, into a visual representation, [3] propose Voice-Driven Type Design (VDTD). It adjusts the shape of each single character according to particular acoustic features in the spoken reference. The motivation of a phoneme-to-grapheme adaptation is to better represent the characteristics of *how* it has been spoken besides *what* has been spoken. VDTD maps the three acoustic properties loudness, speed and pitch to the vertical stroke weight, horizontal stroke weight, and character width. [31] investigate how visual coding of prosody (bolded if louder, squished—what we refer to as narrow—if faster, etc.) can help children improve reading prosody. They found that coding verbal information can create an intuitive representation of speech's expressiveness. [34] exploit variable font technology to visualize voice characteristics at the syllable level and use letter slant to indicate prosody. They report that "participants' responses are highly consistent, indicating that it is indeed plausible to use typographic modulations as a way of representing speech expressiveness, or simply prosody". In another study [35], they report that when an example of their voice-modulated typography was shown along with two alternative sounds, participants correctly identified the original sound with an accuracy of 65%.

To the best of our knowledge, the first technology to visualize voice characteristics in captions is *WaveFont* [4, 5]. *WaveFont* uses methods from automatic speech recognition, signal processing, machine learning, subtitling and typography to render characteristics automatically and intuitively from the voice in captions. In the next section we will motivate our visualization and present our pipeline to generate *WaveFont* captions.

3 AI-Based Visualization of Voice Characteristics in Captions

Our *WaveFont* visualization of the voice characteristics for video captions is at the word level, i.e., for each word, the average values of volume and length are used to decide which font to use to represent the whole word. It was very important for us to present the characteristics of the voice as intuitively as possible so that the viewer does not have to think long about how to interpret the visualization. Our previous analyses have shown that with captions, the word level is preferred to the character or syllable level since viewers see each caption only briefly and therefore have only a short time to interpret the visualization. Furthermore, although a representation of pitch is possible with our technology, we omit it as an additional visualized voice feature in our captions. The reasons are that previous studies have demonstrated that there is no clear agreement about the visualization of pitch in the typeface, pitch plays a subordinate role to the viewers compared to loudness and speed, and we do not want to overwhelm them with information. Consequently, we investigated different mappings and decided to map the voice to the character shape as follows:

- *Loudness*: Producing loudness in speech amplifies the signal and is usually used to attain the attention of a listener. To have the attention of the reader, bolder text is commonly used since it makes it easier and more efficient to scan the text and recognize important keywords [36]. Therefore, we use a thin font for quieter words and a bold font for louder words.
- *Speed*: The processes of information transfer with speech and reading happens within a time period. A reader usually jumps from a part of a word to the next part of a word [37]. Increasing the character width extends this scanning process of the eyes. Thus, we map the speed of the utterance to the character width: We use a narrow font for fast words and a wide font for slow words.

Our mapping is universally understood across cultures, while this is not the case for emojis and emoticons which may convey several meanings [33]. For aesthetic reasons the different fonts are chosen from the same font family. On the one hand, we aim not to have too extreme differences between the fonts, so that the typeface does not look too restless. On the other hand, the fonts need to be different enough to be easily recognized—even by inexperienced viewers on a small screen.

Fig. 1. *WaveFont* captions.

Figure 1 shows excerpts of Martin Luther King Jr.'s speech and of a German lecture video with *WaveFont* captions. The combination of the two visualizations for loudness and speed results in four classes. Figure 2 summarizes the mapping of the acoustic characteristics *loudness* and *speed* to its visual representations *stroke weight* and *character width* in our four classes.

To generate *WaveFont* captions, we use a video and the corresponding caption file as input and apply the following steps [5]:

1. *Extraction of the audio track* from the video.
2. *Segmentation* of the audio track *into smaller audio files* containing spoken utterances based on the time information (start time and end time) of each caption.
3. For each audio speech segment: *Automatic forced alignment* process, which takes the text transcription of the audio speech segment and provides the start and end time of each word in the speech segment.

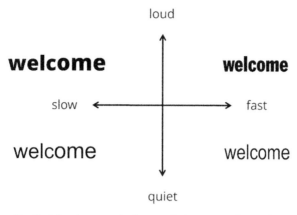

Fig. 2. Mapping speech characteristics on text formatting.

4. *Acoustic feature extraction*: Provides feature values of loudness and speed of each word. Loudness is based on the signal power. The feature values which represent the speed of each spoken word are computed based on the number of characters and the duration of the uttered word.
5. *Mapping of acoustic features to font classes* based on thresholds.
6. *Type design*: Based on the content of the original subtitle file a new subtitle file is produced that contains font definitions according to the mapped font classes.

A detailed technical description is given in [3, 4]. A benefit is that our technology can be ported to new languages and writing systems. For example, in [5], we describe how we adapted the system to Arabic.

4 Experimental Setup

In this section we describe the experimental setup of our study, which we conducted with a questionnaire.

4.1 Study Design

In order to ask questions about the two types of captions, we showed the participants footage from lectures with standard captions and with *WaveFont* captions at the beginning of the questionnaire. To get a representative lecture video for this purpose, we created a 1:46 min compilation of excerpts from 7 lecture videos which were provided by the *Digital4Humanites* project[2]. The selected video snippets meet different criteria in terms of format and content: They are screencasts, slidecasts or animation videos and either address a theoretical treatise, the use of a software or an application. Six of the videos are in German and one in English. Furthermore, they cover various (digital) humanities subject areas: digital image measurement, museum data research, linguistics, 3D digital

---
[2] BMBF funding number: 16DHB3006; running time 1.1.2020–31.12.2022.

reconstruction, and general data-based skills for students. Our video compilation consists of an upper part where the lectures' video snippets are shown with standard captions and a lower part where the same snippets are shown simultaneously with *WaveFont* captions. This has the following advantages: (1) In contrast to having two separate videos, we exclude that participants watch the video with standard captions longer than the video with *WaveFont* captions or vice versa, which could influence the feedback. (2) The participants do not have to watch our video compilation twice, which would prolong the participation in our survey and could possibly lead to a decrease in the participants' motivation. (3) We enable a direct comparison of both visualizations.

In order to guarantee the participants the sole focus on the captions' visualization and a fair comparison, it was important to us that the standard captions and the *WaveFont* captions contain the same captions in terms of content and that these were created according to the best possible subtitle standards. Consequently, when we created the captions, we sticked to the German subtitle standards from ARD and ZDF[3] for the captions of the German lecture videos and to the BBC Subtitle Guidelines[4] for the English lecture video.

Our questionnaire was provided in German and English and consisted of a section where we asked for socio-demographic data, a section with general questions about the use of captions, and a section where we asked participants to compare standard and *WaveFont* captions in the categories *visualization of voice characteristics, understanding the content, following the content, linguistic understanding,* and *identifying important words*. The participants evaluated the questions with a score. The score range follows the rules of a forced choice Likert scale, which ranges from (1) *strongly disagree* to (5) *strongly agree*.

4.2 Participants

48 participants (23 female, 23 male, 2 diverse) filled out our questionnaire. The participants were students or former students at public or private universities and technical colleges, most of whom were between 18 and 44 years old and all participated free of charge. They represent students from a variety of disciplines, such as law, economics and social sciences, engineering, humanities, design and art as well as mathematics and natural sciences. Most of them have a high proficiency in the German language. But some participants are foreign students who do not speak German that well but are enrolled in German universities and listen to German lectures. 50% accessed our questionnaire via laptop or PC, 50% with their smartphones. We appreciate these distributions as it was important to us to get feedback from different people representing the target group who watch lecture videos.

5 Experiments and Results

In this section, we will describe the results of our study in which we compared standard captions to *WaveFont* captions with regard to the categories *visualization of voice characteristics, understanding the content, following the content, linguistic understanding,*

---
[3] http://www.untertitelrichtlinien.de.
[4] https://bbc.github.io/subtitle-guidelines.

and *identifying important words*. In addition, we will investigate for which applications standard and *WaveFont* captions have potential.

5.1 Visualization of Voice Characteristics

After the participants watched our 1:46 min compilation of lecture videos, we asked them in our questionnaire if standard captions (*standard*) or *WaveFont* captions (*WaveFont*) visualize the characteristics of the lecturer's voice (e.g., loudness, lengths, pauses) better for them. Figure 3 illustrates their feedback. The blue pieces in the pie chart represent the proportion of participants who find *standard* better for this category. The green pieces represent the proportion who prefer *WaveFont*. As demonstrated in the figure, exactly half of the participants state that *WaveFont* shows them the characteristics of the voice *much better*. In addition, 27% indicate that *WaveFont* demonstrates these properties *better*. 15% let us know that *standard* visualizes the voice characteristics *better* and 0% that it shows these characteristics *much better*. 8% did not indicate a favorite method. Totally, in this category, 63% are more convinced of *WaveFont* than of *standard*.

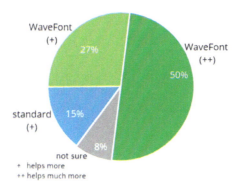

Fig. 3. Visualization of voice characteristics.

To get a better understanding if participants understand the concept of *WaveFont*, we asked if the participants find the visualization of *loudness*, the visualization of *speed* and finally the joint visualization of loudness and speed (*loudness + speed*) comprehensible. As illustrated in Fig. 4, the visualization of *loudness* (3.90) was very well understood. Participants had more trouble understanding the visualization of *speed* (3.54), which is above the average score, indicating that it is still an adequate representation. The joint representation (*loudness + speed*) reduced comprehension (3.38).

However, when we asked if the participants agree that they assume to understand *WaveFont* with more practice even better, the majority agrees with an average of 4.19 in our 5-Likert scale as shown in Fig. 5.

In free text fields of the questionnaire, 4 students suggested using emojis and colors for the visualization of voice characteristics as well as including the lecturer's talking head (picture-in-picture). We do not consider these suggestions useful for the following reasons: While the representation of loudness with the stroke weight and of speed with

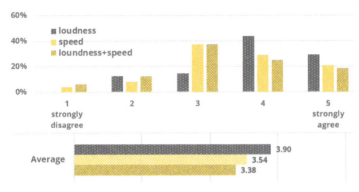

Fig. 4. Comprehensibility of *loudness*, *speed*, and *loudness + speed*.

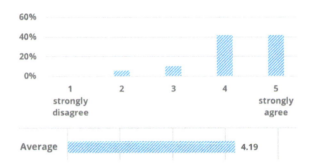

Fig. 5. Assumption of improved *WaveFont* comprehension with more practice.

the character width is intuitive [36, 37], the interpretation of emojis depends on the cultural backgrounds, context, and individual characteristics [28]. The use of different colors in captions and subtitles already indicates different speakers in the BBC Subtitle Guidelines for English. In addition, the lecturer's talking head cannot always be faded in, e.g., not if it obscures important visual components or if the screen is too small.

5.2 Preferences

Figure 6 shows the participants' preferences with regard to captioning in lecture videos. For most students, *WaveFont* is the first choice. *Standard* is chosen as second priority and no captions as third priority. The fact that more students prefer captions than no captions confirms the findings on educational videos described in [1]. The fact that the students favor *WaveFont*, which is more differentiated compared to *standard*, indicates that they like the advantages of visualizing loudness and speed using the typeface. The results of our more detailed analysis of acceptance are described in Sects. 5.3–5.6.

In addition to the general preference, we asked under which conditions the students find the use of *standard* or *WaveFont* in lecture videos particularly important. We received feedback that *standard* and *WaveFont* are particularly preferred when the sound quality of the video is poor, when there is background noise, when the lecturer speaks a language

Fig. 6. Preferences in terms of captions in lecture videos (1st, 2nd, 3rd priority).

of which the students are not native speakers, and when the lecturer has a poor speaking style.

5.3 Understanding the Content

Since in lectures it is important that students understand the content as best as possible, we asked the participants if *standard* or *WaveFont* helps them *more* or even *much more* to understand the general content of the lecture video and specifically what the lecturer is saying in terms of content.

As shown in Fig. 7 (a), 25% reported that *standard* helps them *more* with the general content understanding and 6% indicated that *standard* helps them even *much more*. In contrast, 27% ticked off that *WaveFont* helps them with understanding the content *more* than *standard* and even 29% that *WaveFont* helps them *much more*. 13% were not sure. In total, *WaveFont* convinced 56% of the participants in this category which is 25% more than those who think that *standard* is more helpful in this category.

As visualized in Fig. 7 (b), more participants (21%) were not sure about the second question. But again, the majority (21% + 29%) voted that *WaveFont* helps them *more* or *much more* than *standard* to better understand what the lecturer is saying in terms of content which is 21% more than those who voted for *standard*.

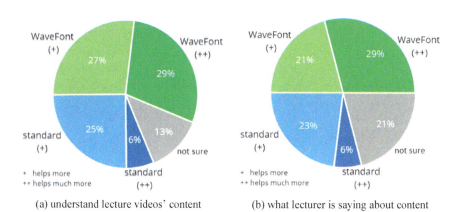

(a) understand lecture videos' content   (b) what lecturer is saying about content

Fig. 7. Content understanding.

5.4 Linguistic Understanding

In addition to content understanding, it was important for us to find out which method helps the students more in terms of linguistic understanding. We wanted to figure out whether *standard* and *WaveFont* help students better grasp linguistic aspects of the lecturer's spoken language, such as grammar and pronunciation. Figure 8 demonstrates that in this category even more welcome *WaveFont* with 63%. The support provided by the visualization of the voice characteristics is particularly evident in the fact that 38%—i.e., 9% more than with *content understanding*—report that *WaveFont* helps them *much more*. Only 27% (23% + 4%) report that *standard* helps them *more* and *much more*. 10% choose neither *standard* nor *WaveFont*.

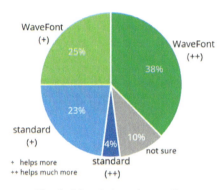

Fig. 8. Linguistic understanding.

5.5 Following the Content

Figure 9 illustrates our evaluation of whether *standard* or *WaveFont* helps them *more* or even *much more* to follow the content of the lecture. This time, slightly more participants reported that *standard* helps them follow the content *more* (27%) and *much more* (10%) than *WaveFont*. However, 27% indicated that *WaveFont* helps them with understanding the content *more* than *standard* and the same percentage that *WaveFont* helps them *much more*. 8% were not sure. Again, *WaveFont* convinced with 54% more than half of the participants in this category which is 17% more than those who prefer *standard* in this category.

5.6 Identifying Important Words

Furthermore, we investigated if our AI-driven *WaveFont* technology helps students better recognize words which are important to the lecturer. As illustrated in Fig. 10, the support provided by the visualization of the speaker's voice characteristics is particularly evident in the large majority of 83% reporting that *WaveFont* helps them *more* (27%) and *much more* (56%). Only 10% report that *standard* helps them *more* and 0% that it helps *much more*. Only 6% were not sure. In this category *WaveFont* outperforms *standard* by 73% which is significantly better than in the other categories.

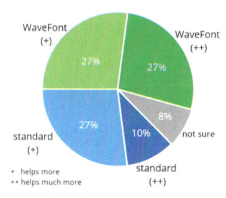

Fig. 9. Following the content.

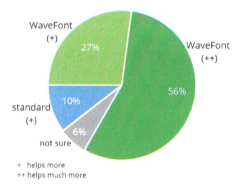

Fig. 10. Identifying important words.

5.7 Further Applications

Finally, we wanted to find out in which other applications the use of *WaveFont* has potential. When asked where the participants would like to see *WaveFont*, we got different answers as visualized in Fig. 11. Use cases where more than 30% of the participants agree are: Live broadcasts, social media, video-on-demand, and videos on websites. In a past study in Arab countries, we also asked this question [5]. In that study, more than 30% of the participants agreed on the following use cases: Video-on-demand, TV, social media, live broadcasts, and TV sets in public places. This shows that there are similarities in the desired use cases between a group of people who are oriented towards German culture and a group of people who are more oriented towards Arabic culture, but also cultural differences.

6 Conclusion and Future Work

In this paper, we have demonstrated that an AI-based visualization of voice characteristics in captions helps students improve the viewing and learning experience in lecture

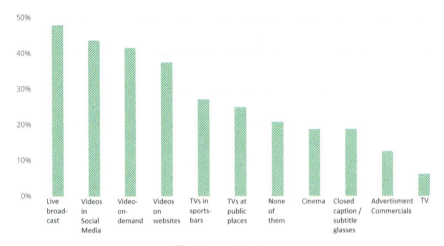

Fig. 11. Applications.

videos. For the AI-based visualization of the lecturer's voice characteristics in captions, we used our *WaveFont* technology [3–5], which processes the speech signal and intuitively displays loudness, speed, and pauses in the subtitle font. In our survey, the AI-based visualization of speech features outperformed standard captions since it helps students visualize speech features, understand the content, follow the content, in linguistic understanding, and identify important words. When we asked if the participants agree that they assume to understand *WaveFont* with more practice even better, the majority agrees with an average of 4.19 in our 5-Likert scale.

Based on this good prediction, we would like to analyze the learning effect in future work in more detail, e.g., with time measurements, measuring cognitive load, eye tracking and with targeted questions about the content of video lectures. Thereby, we also plan to investigate the effect of language acquisition for non-native speakers. Moreover, it is interesting to analyze the effect of *WaveFont* captions on learning styles. On the one hand, visual learners are good at using vision to obtain information, and their information processing channels tend to use visual channels and are more inclined to use pictures to represent information and thoughts. Verbal learners, on the other hand, are good at using auditory information to obtain information, and their information processing channels are more likely to use auditory channels and more inclined to use words to present information and thoughts [39].

References

1. Gernsbacher, M.A.: Video captions benefit everyone. Policy Insights Behav. Brain Sci. **2**(1), 195–202 (2015)
2. Marteney, J.: Verbal and nonverbal communication. ASCCC Open Educational Resources Initiative (OERI). https://socialsci.libretexts.org/@go/page/67152 (2020)
3. Wölfel, M., Schlippe, T., Stitz, A.: Voice driven type design. In: International Conference on Speech Technology and Human-Computer Dialog (SpeD), Bucharest, Romania (2015)

4. Schlippe, T., Wölfel, M., Stitz, A.: Generation of text from an audio speech signal. US Patent 10043519B2 (2018)
5. Schlippe, T., Alessai, S., El-Taweel, G., Wölfel, M., Zaghouani, W.: Visualizing voice characteristics with type design in closed captions for Arabic, International Conference on Cyberworlds (CW 2020), Caen, France (2020)
6. United Nations: Sustainable Development Goals: 17 goals to transform our world. https://www.un.org/sustainabledevelopment/sustainable-development-goals (2021)
7. Correia, A.P., Liu, C., Xu, F.: Evaluating videoconferencing systems for the quality of the educational experience. Distance Educ. **41**(4), 429–452 (2020)
8. Koravuna, S., Surepally, U.K.: Educational gamification and artificial intelligence for promoting digital literacy. Association for Computing Machinery, New York, NY, USA (2020)
9. Chen, L., Chen, P., Lin, Z.: Artificial intelligence in education: A review. IEEE Access **8**, 75264–75278 (2020). https://doi.org/10.1109/ACCESS.2020.2988510
10. Rakhmanov, O., Schlippe, T.: Sentiment analysis for Hausa: Classifying students' comments. The 1st Annual Meeting of the ELRA/ISCA Special Interest Group on Under-Resourced Languages (SIGUL 2022). Marseille, France (2022)
11. Libbrecht, P., Declerck, T., Schlippe, T., Mandl, T., Schiffner, D.: NLP for student and teacher: Concept for an AI based information literacy tutoring system. In: The 29th ACM International Conference on Information and Knowledge Management (CIKM2020). Galway, Ireland (2020)
12. Sawatzki, J., Schlippe, T., Benner-Wickner, M.: Deep learning techniques for automatic short answer grading: Predicting scores for English and German answers. In: The 2nd International Conference on Artificial Intelligence in Education Technology (AIET 2021). Wuhan, China (2021)
13. Schlippe, T., Sawatzki, J.: Cross-lingual automatic short answer grading. In: The 2nd International Conference on Artificial Intelligence in Education Technology (AIET 2021). Wuhan, China (2021)
14. Bothmer, K., Schlippe, T.: Investigating natural language processing techniques for a recommendation system to support employers, job seekers and educational institutions. In: The 23rd International Conference on Artificial Intelligence in Education (AIED) (2022)
15. Bothmer, K., Schlippe, T.: Skill Scanner: Connecting and supporting employers, job seekers and educational institutions with an AI-based recommendation system. In: Proceedings of The Learning Ideas Conference 2022 (15th Annual Conference), New York, 15–17 June (2022)
16. Schlippe, T., Sawatzki, J.: AI-based multilingual interactive exam preparation. In: Guralnick, D., Auer, M.E., Poce, A. (eds.) TLIC 2021. LNNS, vol. 349, pp. 396–408. Springer, Cham (2022). https://doi.org/10.1007/978-3-030-90677-1_38
17. Wölfel, M.: Towards the automatic generation of pedagogical conversational agents from lecture slides. In: International Conference on Multimedia Technology and Enhanced Learning (2021)
18. Ou, C., Joyner, D.A., Goel, A.K.: Designing and developing video lessons for online learning: A seven-principle model. Online Learn. **23**(2), 82–104 (2019)
19. Wang, J., Antonenko, P., Dawson, K.: Does visual attention to the instructor in online video affect learning and learner perceptions? An eye-tracking analysis. Comput. Educ. 146 (2020). https://doi.org/10.1016/j.compedu.2019.103779
20. Perego, E., Del Missier, F., Porta, M., Mosconi, M.: The cognitive effectiveness of subtitle processing. Media Psychol. **13**, 243–272 (2010)
21. Linebarger, D.L.: Learning to read from television: The effects of using captions and narration. J. Educ. Psychol. **93**, 288–298 (2001)
22. Bowe, F.G., Kaufman, A.: Captioned Media: Teacher Perceptions of Potential Value for Students with No Hearing Impairments: A National Survey of Special Educators. Described and Captioned Media Program, Spartanburg, SC (2001)

23. Guo, P.J., Kim, J., Rubin, R.: How video production affects student engagement: An empirical study of MOOC videos. In: L@S'14: Proceedings of the First ACM Conference on Learning. March 2014, pp. 41–50 (2014). https://doi.org/10.1145/2556325.2566239
24. Alfayez, Z.H.: Designing educational videos for university websites based on students' preferences. Online Learn. **25**(2), 280–298 (2021)
25. Persson, J.R., Wattengård, E., Lilledahl, M.B.: The effect of captions and written text on viewing behavior in educational videos. Int. J. Math Sci. Technol. Educ. **7**(1), 124–147 (2019)
26. Vy, Q.V., Fels, D.I.: Using placement and name for speaker identification in captioning. In: Miesenberger, K., Klaus, J., Zagler, W., Karshmer, A. (eds.) ICCHP 2010. LNCS, vol. 6179, pp. 247–254. Springer, Heidelberg (2010). https://doi.org/10.1007/978-3-642-14097-6_40
27. Brown, A., et al.: Dynamic subtitles: The user experience. In: TVX (2015)
28. Fox, W.: Integrated titles: An improved viewing experience. In: Eyetracking and Applied Linguistics (2016)
29. Ohene-Djan, J., Wright, J., Combie-Smith, K.: Emotional subtitles: A system and potential applications for deaf and hearing impaired people. In: CVHI (2007)
30. Rashid, R., Aitken, J., Fels, D.: Expressing emotions using animated text captions. Web Design for Dyslexics: Accessibility of Arabic Content (2006)
31. Bessemans, A., Renckens, M., Bormans, K., Nuyts, E., Larson, K.: Visual prosody supports reading aloud expressively. Visible Lang. **53**, 28–49 (2019)
32. Gernsbacher, M.: Video captions benefit everyone. Policy Insights Behav. Brain Sci. **2**, 195–202 (2015)
33. El-Taweel, G.: Conveying emotions in Arabic SDH: The case of pride and prejudice. Master thesis, Hamad Bin Khalifa University (2016)
34. de Lacerda Pataca, C., Costa, P.D.P.: Speech modulated typography: Towards an affective representation model. In: International Conference on Intelligent User Interfaces, pp. 139–143 (2020)
35. de Lacerda Pataca, C., Dornhofer Paro Costa, P.: Hidden bawls, whispers, and yelps: Can text be made to sound more than just its words? (2022). arXiv:2202.10631
36. Bringhurst, R.: The elements of typographic style, vol. 3.2, pp. 55–56. Hartley and Marks Publishers (2008)
37. Unger, G.: Wie man's liest, pp. 63–65. Niggli Verlag (2006)
38. Bai, Q., Dan, Q., Mu, Z., Yang, M.: A systematic review of emoji: Current research and future perspectives. Front. Psychol. **10**, 2221 (2019). https://doi.org/10.3389/fpsyg.2019.02221
39. Rayner, S.G.: Cognitive styles and learning styles. In: Wright, J.D. (ed.) International Encyclopedia of Social and Behavioral Sciences, vol. 4, 2nd edn, pp. 110–117. Elsevier, Oxford (2015)

The Intergroup Bias in the Effects of Facial Feedback on the Recognition of Micro-expressions

Kunling Peng[1,2], Yaohan Wang[1,2], and Qi Wu[1,2(✉)]

[1] Department of Psychology, School of Educational Sciences, Hunan Normal University, Changsha 410006, Hunan, China
pengkunling@hunnu.edu.cn, Wangyp981115@163.com, sandwich624@yeah.net

[2] Cognition and Human Behavior Key Laboratory of Hunan Province, Hunan Normal University, Changsha 410006, Hunan, China

Abstract. Micro-expression is the facial expression that is extremely quick and lasts less than half a second. As a spontaneous expression, it is usually produced when people try to suppress their emotions and can reveal the true emotions of human beings. It plays an important role in lie detection. In recent years, with the progress of neural network technology, the research on micro-expression recognition has made significant progress. However, because our understanding of the psychological process of micro-expression recognition is far from complete, the existing method of recognizing micro-expression still cannot meet the standard of practical application. In the present research, we investigated the effects of facial feedback and social identity of the expresser on the recognition of micro-expressions by one behavioral experiment. The results showed that facial feedback can moderate the intergroup bias in micro-expression recognition, which suggests that humans will imitate other people's facial expressions to different degrees in the recognition of micro-expression. At shorter duration, facial feedback has a stronger effect on the recognition of micro-expression of outgroup members. And at longer duration, facial feedback has a stronger effect on the recognition of micro-expressions of ingroup members. This further suggests that we need to consider the identity of the model and the identity of the coder to obtain more accurate and effective data coding when establishing the micro-expression database.

Keywords: Micro-expression recognition · Micro-expression database · Facial feedback · Intergroup bias effect · Micro-expression duration

1 Introduction

From the perspective of natural selection, the functions of human emotion are of great significance to our survival and reproduction [1]. Facial expression, as the main symbol used by humans to express their internal feelings and emotions, is the focus of many psychological researchers. Up to now, research on facial expression has been abundant.

At present, researchers believe that some expressions are universal across cultures and have their own unique facial features and specific patterns of brain activity that express specific feelings and motivations [2]. For example, angry expression often means aggression and rejection towards others, while happy expression means that the individual is feeling pleasure. However, not all our feelings are directly expressed by the face. Sometimes human beings need to hide or suppress our true emotions, which results in a very special form of facial expression. That is the micro-expression [3].

1.1 Micro-expression and Micro-expression Recognition

Micro-expression. Micro-expression is usually characterized by short duration (last less than 0.5 s). Different from typical facial expression (also can be referred as macro-expression), it cannot be voluntarily controlled and can express the real emotion that humans intentionally suppress. Therefore, micro-expression can be used to judge whether people are lying [3]. Specifically, the accurate recognition of micro-expression and understanding the relationship between micro-expressions and emotions can provide us clues to infer the real human emotions. Due to the special characteristics of micro-expressions, the study of micro-expression recognition has great application value in medical treatment, national security, communication and negotiation, educational evaluation and other emotion-related fields [4].

Micro-expression Recognition. Micro-expression recognition refers to that the recognizer can accurately distinguish and understand the specific meaning of the micro-expression. Facial expression recognition (FER) is a mature field and many efficient algorithms for FER have been developed [5]. However, identifying the micro-expression is very difficult due to its own characteristics, such as short presentation time and only partial facial muscle movement [6]. In previous researches, psychologists have developed tests that can screen individuals' ability to recognize micro-expression and are trying to develop training tools that can help those with higher talent for micro-expression recognition [7]. In 1974, Ekman and Friesen developed a program called the Brief Affect Recognition Test (BART) [8]. In the following years, other researchers continued to improve this test. For example, the famous Japanese and Caucasian Brief Affect Recognition Test (JACBART) developed by Matsumoto and Ekman provides high ecological validity by presenting dynamic image [9]. It was not until Ekman developed Micro Expression Training Tool (METT) in 2002 that significant progress was made in the field of micro-expression recognition training [10]. The training effect of METT is quite good: after only 90 min of training, the recognizers' post-test scores are 30%~40% higher on average than the pre-test, and the recognition ability of the recognizers has been significantly improved [10]. Frank found that the program was effective in training people from different backgrounds [11]. In 2011, Matsumoto and his colleague jointly developed a new micro-expression training program based on METT, which can effectively improve the accuracy of the participants' micro-expression recognition and promote their social communication ability. And the results showed that the effect of this training can be maintained in a long time. This demonstrates the effectiveness of micro-expression recognition training [12]. However, although previous researches seem to suggest that such training tools are effective, Frank's research results showed that even after METT training, the accuracy of recognizing spontaneous micro-expressions was still less than 40%, which indicated that the tool is not efficient enough in practice [11].

Because of the great application value of micro-expression, computer scientists also pay great attention to the automatic recognition of micro-expression. The development of computer vision and video processing technology has gradually made the development of such automatic tools possible [4]. While psychologists and computer scientists have each made some progress, the accuracy of micro-expression recognition is still not up to the standard of practical application [13]. Exploring the influencing factors of micro-expression recognition can help us to solve this problem. One of these important factors that may affect the recognition of micro-expression is the facial feedback signal. It often occurs spontaneously in the process of micro-expression recognition [14].

1.2 Facial Feedback

Previous studies have found that people's imitation of facial expressions has a feedback effect on their subjective experience of real emotions [14]. Specifically, studies found that people's imitation of facial expressions can not only regulate the subjective emotional experience generated by the facial muscle action, but also change the intensity of the subjective emotional experience induced by stimulus situation [15]. For example, if one is happy, the more he laughs, the happier he will be. On the contrary, if he shows the opposite expression of happiness (such as scowl), then the degree of happiness will be reduced [15].

Since facial feedback can regulate the emotions of observers, it will inevitably affect the process of facial expression recognition. From the perspective of embodied cognition, the observers do not judge other people's emotions directly by observing their facial expressions, but partly by imitating the perceived expressions with his own facial muscles. In this regard, embodied cognition theory holds that the process of facial feedback to help facial expression recognition is as follows: First, the observers unconsciously imitate the target facial expression; Second, a subtle muscle contraction in the face generates an afferent muscle feedback signal that goes to the brain; Third, the observers use this facial feedback to reproduce the observed expression, thus identifying the emotional meaning of the expression [16].

How does facial feedback affect micro-expression recognition? Previous studies have shown that facial feedback could be a contributing internal cue for the recognition and judgment of micro-expression. For high-intensity micro-expression, the suppression of facial feedback did improve the accuracy of subjects' recognition; for low-intensity micro-expression, the enhancement of facial feedback reduced the accuracy [17, 18]. The results indicated that the manipulation of facial feedback has a detrimental effect on the recognition of micro-expression. Moreover, researchers found that this detrimental effect mainly comes from the lower face of micro-expressions [18].

1.3 Intergroup Bias

In addition to the influence of facial feedback, the researchers also found that the social identity of expresser has an important influence on the recognition of macro-expression and micro-expression. For macro-expression, Elfenbein and Ambady found in their study that there was an intergroup bias effect in human facial emotion recognition. That is, when people use facial expressions to judge the emotions of people from the same culture as themselves, the accuracy rate is higher. When it comes to judge the emotions of people from different cultural backgrounds, the accuracy is lower [19]. In 2004, Elfenbein et al. found that regardless of whether the cultural background is the same or not, the accuracy of emotion recognition on the right side of the face is significantly higher than that on the left half of the face. In other words, the intergroup bias effect is more significant in the left half of the face [20]. In addition, Wood and Rychlowska concluded in their review that in the context of western culture, participants mimicked the expressions of ingroup members to a greater extent than those of outgroup members [21]. For micro-expression, Xie et al. found that under the background of eastern culture, the accuracy of Chinese participants in the recognition of White models' micro-expressions was higher than that of Asian models' micro-expressions regardless of the duration of micro-expressions. This effect even persisted after the training of METT [22]. These results indicated that, contrary to the ingroup advantage found in the macro-expression recognition, participants actually seem to harbor an ingroup disadvantage (also be referred as ingroup derogation) in micro-expression recognition [22]. This intergroup bias effect in micro-expression recognition deserves further investigation.

2 Overview of the Current Research

As mentioned above, the researchers found that facial feedback is an inhibitory cue that is "harmful" to micro-expression recognition, and its effect is moderated by the intensity of micro-expression [17]. Previous studies have also shown that ethnic group is a universal external cue for identifying individuals as same group. Individuals usually regard people from the same ethnic group of themselves as ingroup, while people of different ethnic group as the outgroup [22, 23]. So, does facial feedback modulate the effect of social identity on micro-expression recognition? Previous studies have found that for macro-expressions, participants will imitate the facial actions of their ingroup members to a greater extent than that of outgroup members [21]. This suggests that enhancing the facial feedback will increase the ingroup disadvantage for micro-expression recognition. In the present study, we used the ethnic group to manipulate the group identity. Since the subjects were Chinese students, Asian faces were used as the ingroup faces, while the White faces were used as the outgroup faces. The Black faces were not employed in order to avoid the potential interference from the stereotypes. Specifically, we investigated whether enhanced facial feedback has different effects on the recognition accuracy of micro-expressions of different social identity of expresser. In addition, we also investigated that intergroup bias of facial feedback on the recognition on micro-expression presented at different durations.

3 Method

3.1 Participants

We used the Gpower software (V3.1; Faul et al. 2009) to estimate our sample size [24]. By using the parameters (medium effect size: $f = 0.25$) and given the current experimental design, this analysis suggested we should recruit at least 128 participants to obtain the power of 0.8. Consequently, a total of 128 Chinese college students ($M_{age} = 19.91$, $SD = 1.14$, 92 females) were recruited from Hunan Normal University. Participants had normal or corrected-to-normal vision. All participants gave informed consent and were given partial course credit or monetary reward after the experiment.

3.2 Design

A 2 (facial feedback: enhanced, control) × 2 (target group identity: ingroup, outgroup) × 2 (micro-expression duration: 150 ms, 450 ms) mixed-model experimental design was used, with facial feedback being the between-subjects factor while target group identity and micro-expression duration being the within-subjects factors. Dependent variables were the accuracy of micro-expression recognition, accuracy of working memory task, and response time of working memory task.

3.3 Stimuli

Twelve models (six Whites, six Asians; half male and half female) from the BU-3DFE database were selected [25]. For each of these selected models, the images of his/her seven facial expressions (i.e., sadness, surprise, anger, disgust, fear, happiness and neutral) were selected as the stimulus materials. Therefore, a total of 84 facial images were selected and the size of each image was 512 × 512 pixels. All facial images were converted to gray-scale images and their gray values were normalized to grand mean. In previous research, these processed facial expressions images of the six White and Asian models have been validated that their recognition accuracy were matched when they were presented as macro-expressions [22]. The combinations of the selected models and the duration conditions (i.e., 150 ms and 450 ms) were completely random and even. During the experiment, six universal facial expressions of the selected models were presented randomly as target stimuli. Since these expressions were presented for less than 0.5 s, they could be regarded as micro-expressions.

3.4 Manipulation of Facial Feedback

For the manipulation and control of facial feedback, we adopted the method used by Neal and Chartrand [26]. Subjects in the enhancement group applied a gel composed of polyvinyl alcohol evenly over their entire face, while subjects in the control group applied the gel evenly over the inside of their non-dominant arm. All subjects were required to keep applying gel until the end of the experiment. In addition, to test the effectiveness of facial feedback manipulation, all participants in the enhancement group were required to fill out an impedance questionnaire. That is, the sixty-four subjects in the enhancement group were asked to exercise their facial muscles twice (5 min before and 10 min after gel application) and to report the degree of resistance (1 = no resistance at all; 11 = strong impedance) [26].

3.5 Experimental Tasks

The subjects were randomly divided into two groups. In the experiment, subjects in each group were required to complete two experimental tasks, the subjects first completed the micro-expression recognition task, and then completed the working memory task.

Micro-expression Recognition Task. We adopted JACBART paradigm for micro-expression presentation (See Fig. 1) [9].

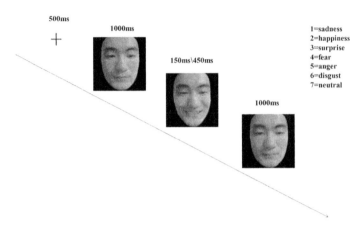

Fig. 1. Experimental procedures

First, a "+" was shown in the center of a computer monitor screen for 500 ms, followed by a neutral expression of a model in the center of the screen for 1000 ms, and then one of the model's six basic expressions was quickly shown at the same location. Duration was one of two levels (150 ms, 450 ms). After the basic expression was displayed, the neutral expression picture was displayed again in the center of the screen for 1000 ms. In each trial, the neutral expressions and basic expressions were taken from the same model. Then the subjects were asked to choose one of the seven emoticon labels of sadness, surprise, anger, disgust, fear, happiness or neutral to judge the emotion just flashed on the screen (corresponding to the buttons from 1 to 7), with no limit on the time. These stimuli were presented in a completely randomized order.

Working Memory Task. Different locations of gel application may bring different degrees of cognitive load, and the change of cognitive load may affect micro-expression recognition [27]. In order to rule out the possibility that the effect of facial feedback manipulation was caused by changing the cognitive load, a working memory task was set up. In the task, the subjects were required to quickly judge whether the results of 16 modular mathematical operations were correct. The task is highly sensitive to changes in cognitive load [9].

4 Results

First of all, the result of pair-wise t-test showed that the degree of impedance to facial muscles in the enhanced group was significantly increased after gel application ($t(63) =$

−33.58, $p < 0.001$ (before gel application: $M = 1.16$, $SD = 0.44$; after gel application: $M = 8.03$, $SD = 1.60$). This result indicated that the facial feedback manipulation was effective.

We conducted a 2 (facial feedback: enhanced and control) × 2 (target group identity: ingroup and outgroup) × 2 (micro-expression duration: 150 ms and 450ms) analysis of variance (ANOVA) on recognition accuracy. This result indicated that the main effect of the target group identity was significant ($F (1, 126) = 11.51$, $p = 0.001$, partial $\eta^2 = 0.08$). The participants were more accurate in recognizing the micro-expression of the outgroup members than the ingroup members. The main effect of micro-expression duration was significant ($F (1, 126) = 38.29$, $p < 0.001$, partial $\eta^2 = 0.23$). The accuracy of the micro-expression recognition was higher under the 450ms condition than under the 150ms condition. The main effect of facial feedback was not significant ($F (1, 126) = 0.76$, $p = 0.387$, partial $\eta^2 = 0.01$).

The interaction among the three factors (target group identity, micro-expression duration and facial feedback) was significant ($F (1, 126) = 4.26$, $p = 0.041$, partial $\eta^2 = 0.03$). Further simple effects analysis showed that enhanced facial feedback exaggerated or inhibited the ingroup disadvantage of micro-expression recognition under certain conditions. Under the condition of 150ms, there was no significant difference in the accuracy of the enhanced group for the ingroup and outgroup members' micro-expression recognition ($F (1, 126) = 0.11$, $p = 0.742$, partial $\eta^2 = 0.001$), while in the control group, participants were more likely to recognize the micro-expression of outgroup members ($F (1, 126) = 9.85$, $p = 0.002$, partial $\eta^2 = 0.07$). Under the condition of 450 ms, there were no significant differences between the recognition accuracy of the micro-expressions of ingroup members and the recognition accuracy of the micro-expressions of outgroup member in the control group ($F (1, 126) = 1.69$, $p = 0.196$, partial $\eta^2 = 0.01$). But, in the enhanced group, participants were more likely to recognize the micro-expression of outgroup members ($F (1, 126) = 5.63$, $p = 0.019$, partial $\eta^2 = 0.04$) (See Fig. 2). The results indicate that enhanced facial feedback has different effects on micro-expression recognition of different target groups depending on the duration of micro-expressions.

In addition, independent sample t-tests were performed on the data of working memory task, and the results showed that there were no significant differences between the control and enhanced group in the accuracy (control group: $M = 0.65$, $SD = 0.17$; enhanced group: $M = 0.66$, $SD = 0.19$; $t (126) = -0.15$, $p = 0.88$) and in the reaction time (control group: $M = 4880.45$ ms, $SD = 1269.60$; enhanced group: $M = 4901.79$ ms, $SD = 1622.42$; $t (126) = -0.08$, $p = 0.93$). These results indicate that the results found in the micro-expression recognition task cannot be attributed to the differences caused by facial feedback manipulations.

5 Discussion

In the present study, a behavioral experiment was conducted to investigate the effects of facial feedback and social identity of the expresser on the recognition process of micro-expressions.

The main effect of target group identity was significant, and the accuracy of participants in identifying the micro-expression of outgroup member was higher than that

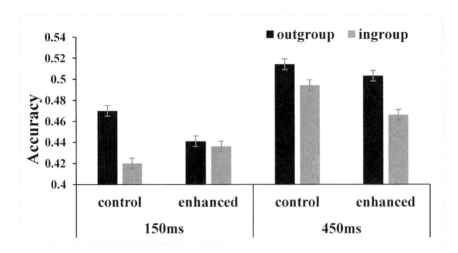

Fig. 2. Mean recognition accuracy of micro-expression under different conditions ($M \pm SE$).

of ingroup member. From the perspective of evolutionary psychology, this may be due to the biological adaptability of ingroup derogation of micro-expression recognition in eastern cultures. The psychological mechanism of ingroup derogation makes individuals more sensitive to the emotions of the outgroup, so that they can adapt to new partners more quickly and cooperate better with outgroup. From the perspective of perceptual processing motivation, the motivation difference of micro-expression recognition leads that the participants had a processing advantage in the perception of outgroup. The brain mechanism was optimized to decode the expression of outgroup, which resulted in the phenomenon of the Chinese participants' ingroup derogation in the recognition of the ingroup and outgroup members' micro-expression. The result indicated that individuals' recognition of different ethnic micro-expression is inconsistent. The present research only employed faces of Whites and Asians in micro-expression recognition, and further experimental investigation is needed to determine whether there are also differences between other ethnic groups. Existing micro-expression databases mostly lack data of different ethnic micro-expression, which makes the micro-expression recognition models trained on these databases less reliable [28]. Therefore, the enrichment of micro-expression samples from different ethnic groups has become a prerequisite for future micro-expression research.

In the present study, we also found that the inhibitory effect of facial feedback on the recognition of micro-expression between ingroup and outgroup was different depending on the duration. Under 150 ms condition, facial feedback had more influence on the recognition of outgroup; under 450 ms condition, facial feedback had more influence on the recognition of ingroup. It has also been found in previous studies that individuals have different recognition mechanisms for expression of different ethnic groups. For expression of outgroup, process of recognition is quick and rough. While for expression of ingroup, process of recognition is slow and precise. When the expressions are presented for a short time, the expressions of outgroup members can capture our attention more easily. When the expressions are presented for a long time, the expressions of ingroup members have a processing advantage [29]. Therefore, the results of the current study indicate that in the process of micro-expression recognition, not only the spatial features need to be analyzed, but also the temporal features of micro-expression need to be used. For example, a new deep learning framework which combines convolutional neural network (CNN) with long short-term memory (LSTM) has certain advantages in extracting spatial and temporal features of expression. This kind of micro-expression analysis technology can be used for real-time micro-expression recognition [30].

6 Conclusion

In summary, the current study found that facial feedback can moderate the intergroup bias in micro-expression recognition, which suggests that humans will imitate other people's facial expressions to different degrees in the recognition of micro-expression.

Acknowledgement. This work was supported by the Outstanding Young Scientific Research Project of Hunan Provincial Department of Education (19B361).

References

1. Laith, A.S., Daniel, C.B., Kelly, A., David, M.B.: Human emotions: an evolutionary psychological perspective. Emot. Rev. **8**(2), 173–186 (2015)
2. Carroll, E.I.: Innate and universal facials expressions: evidence from developmental and cross-culture research. Psychol. Bull. **115**(2), 288–299 (1994)
3. Ekman, P.: Darwin, deception, and facial expression. Ann. N. Y. Acad. Sci. **1000**(1), 205–221 (2006)
4. Zhang, L.-F., Ognjen, A.: Review of automatic micro-expression recognition in the past decade. MAKE **3**(2), 414–434 (2021)
5. Olufisayo, S.E., Serestina, V.: Facial expression recognition: a review of trends and techniques. IEEE Access **9**, 136944–136973 (2021)
6. Yan, W.-J., Wu, Q., Liang, J., Chen, Y.-H., Fu, X.: How fast are the leaked facial expressions: the duration of micro-expressions. J. Nonverbal Behav. **37**(4), 217–230 (2013). https://doi.org/10.1007/s10919-013-0159-8
7. Porter, S., Ten, B.L.: Reading between the lies: identifying concealed and falsified emotions in universal facial expressions. Psychol. Sci. **19**(5), 508–514 (2008)
8. Ekman, P., Friesen, W.V.: Detecting deception from the body or face. J. Pers. Soc. Psychol. **29**(3), 288–298 (1974)
9. Matsumoto, D., Leroux, J., Wilsoncohn, C.: A new test to measure emotion recognition ability: Matsumoto and Ekman's Japanese and Caucasian Brief Affect Recognition Test (JACBERT). J. Nonverbal Behav. **24**(3), 179–209 (2000)
10. Ekman, P.: Microexpression Training Tool [EB]. http://www.paulekman.com. Accessed 15 April 2009
11. Frank, M.G., Herbasz, M., Sinuk, K., Keller, A., Nolan, C.: I see how you feel: training laypeople and professionals to recognize fleeting emotions. In: The Annual Meeting of the International Communication Association. http://www.allacademic.com/meta/p15018_index.html. Accessed 1 July 2009
12. Matsumoto, D., Hwang, H.S.: Evidence for training the ability to read microexpressions of emotion. Motiv. Emot. **35**(2), 181–191 (2011)

13. Kang, J., Chen, X.-Y., Liu, Q.-Y., Jin, S.-H., Yang, C.-H., Hu, C.: Research on a microexpression recognition technology based on multimodal fusion. Complexity **2021**, 1–15 (2021)
14. Kathleen, R.B., David, M.: Facial mimicry is not necessary to recognize emotion: facial expression recognition by people with Moebius syndrome. Soc. Neurosci. **5**(2), 241–251 (2010)
15. William, F.: Peripheral feedback effects of facial expressions, bodily postures, and vocal expressions on emotional feelings. Cogn. Emot. **20**(2), 177–195 (2006)
16. Wang, J.-J.: The Effect of Facial Feedback on Facial Expression Recognition and Its Influence Factor. ZheJiang Normal University (2015) (in Chinese)
17. Wu, Q., Guo, H., He, L.-L.: Facial feedback and micro-expression recognition. J. Psychol. Sci. **39**(6), 1353–1358 (2016). (in Chinese)
18. Zeng, X.-M., Wu, Q., Zhang, S.-W., Liu, Z.-Y., Zhou, Q., Zhang, M.-S.: A false trail to follow: differential effects of the facial feedback signals from the upper and lower face on the recognition of micro-expressions. Front. Psychol. **9** (2015)
19. Elfenbein, H.A., Ambady, N.: Is there an in-group advantage in emotion recognition? Psychol. Bull. **128**(2), 243–249 (2002)
20. Hillary, A.E., Manas, K.M., Nalini, A., Susumu, H., Surender, K.: Hemifacial differences in the in-group advantage in emotion recognition. Cogn. Emot. **18**(5), 613–629 (2004)
21. Gray, H.M., Mendes, W.B., Dennybrown, C.: An in-group advantage in detecting intergroup anxiety. Psychol. Sci. **19**(12), 1233–1237 (2008)
22. Xie, Y.-N., Zhong, C.-Y., Zhang, F.-Q., Wu, Q.: The ingroup disadvantage in the recognition of micro-expressions. In: 14th IEEE International Conference on Automatic Face & Gesture Recognition (FG 2019), pp. 1–5. IEEE, Lille, France (2019)
23. Miller, S.L., Maner, J.K., Becker, D.V.: Self-protective biases in group categorization: threat cues shape the psychological boundary between "us" and "them." J. Pers. Soc. Psychol. **99**(1), 62–77 (2010)
24. Faul, F., Erdfelder, E., Buchner, A., Lang, A.G.: Statistical power analyses using GPower 3.1: tests for correlation and regression analyses. Behav. Res. Methods **41**(4), 1149–1160 (2009)
25. Yin, L., Wei, X., Sun, Y., Wang, J., Rosato, M.J.: A 3D facial expression database for facial behavior research. In: 7th International Conference on Automatic Face and Gesture Recognition (FGR06), pp. 211–216. IEEE, Southampton (2006)
26. Neal, D.T., Chartrand, T.L.: Embodied emotion perception: amplifying and dampening facial feedback modulates emotion perception accuracy. Soc. Psychol. Pers. Sci. **2**(6), 673–678 (2011)
27. Beilock, S.L., Kulp, C.A., Holt, L.E., Carr, T.H.: More on the fragility of performance: choking under pressure in mathematical problem solving. J. Exp. Psychol. **133**(4), 584–600 (2004)
28. Yan, W.-J., Li, X.-B., Wang, S.-J., Zhao, J.-Y., Liu, Y.-J., Chen, Y.-H., Fu, X.-L.: CASME II: an improved spontaneous micro-expression database and the baseline evaluation. PLoS ONE **9**(1), e86041 (2014)
29. Young, S.G., Hugenberg, K.: Mere social categorization modulates identification of facial expressions of emotion. J. Pers. Soc. Psychol. **99**(6), 964–977 (2010)
30. Saranya, R., Poongodi, C., Somasundaram, D., Nirmala, M.: Novel deep learning model for facial expression recognition based on maximum boosted CNN and LSTM. Emot. Rev. **8**(2), 173–186 (2016)

Feedback on the Result of Online Learning of University Students of Health Sciences

Carmen Chauca[✉] [iD], Ynés Phun-Pat [iD], Maritza Arones [iD], and Olga Curro-Urbano [iD]

Universidad Nacional San Luis Gonzaga, Ica, Perú
{carmen.chauca,yphun,marones,ocurro}@unica.edu.pe

Abstract. Faced with the challenge of online teaching-learning, university teachers continued with the responsibility of developing their learning sessions, innovating teaching material and methodology during this process, changing the way of generating learning in health sciences students, through the application of videos, summary readings and practices carried out with family members who acted as patients, in order to achieve the planned competition. The importance of letting students know their achievements in relation to what is evaluated, helps them to understand their way of learning, assess their learning result and self-regulate. This is how feedback motivates the student to rethink their learning strategies. The purpose of this study was to determine the effect of feedback on the online learning outcome of health sciences university students, in a non-experimental research, descriptive-correlational level, with a sample of 294 students. The results obtained showed that feedback in university students of Health Sciences in virtual environments is effective when applied in a timely manner and can be planned, based on the evidence of the learning outcome. To achieve this, they must be previously trained, from the first semesters of study, in feedback literacy, making it part of the self-regulation of their learning.

Keywords: Feedback · Online learning · Online assessment

1 Introduction

In the current context of online learning, activities related to the teaching-learning process have been a permanent challenge for the university teacher, not only in the development of teaching material but also in the innovation of evaluation proposals. Learning-oriented training [1, 2]. During this process, it is necessary for the student to reflect on his cognitive activity [3, 4] that allows him to check the progress of his learning from the beginning to the end [5] of each university academic semester and maintain his active motivation throughout. The process [6, 7]. In the last two years, online education, developed by the Public University of the city of Ica, has experienced great changes in professional careers in Health Sciences, making it necessary to carry out more flexible teaching-learning processes [8]. For example, subjects of a theoretical-practical nature were developed using digital tools [9] such as chats, virtual classroom, institutional emails, video recordings, forums and evaluations that serve to verify the acquisition of knowledge [10–12].

Learning modules as well as online platforms have been implemented with greater emphasis [13, 14] It is very true that the Health Sciences careers, due to the characteristics of their theoretical-practical content and the requirement of the development of the practice in laboratories, clinics, hospitals, are essential requirements in the professional training of the student [15], this allows them to develop not only knowledge, but also abilities and skills [16, 17] to achieve competence in their professional training. Although the educational process in online scenarios "requires varied opportunities for dialogue between the teacher and the student" [18], it made the student suspend the practice [19] and replace it with video analysis of clinical cases, readings of cases, which after observing or reading, had to answer a questionnaire or questions from the teacher. In other cases, he carried out simulations [20] with a closest relative, who acted as a patient, complying with the indications of the practice teacher, questions were asked and the other students had the possibility to appreciate and ask. With this methodology used in the subjects of the professional careers of Health Sciences, it led us to reflect on whether the feedback used by the teacher was appropriate, since it is aimed at reinforcing the development of students' skills [21].

In the study carried out by [22] on feedback in online learning environments, it shows that automatic feedback increases student performance in activities by 65% and automatic feedback is more efficient than manual feedback by 82%. Also [23] studied whether feedback strategies in formative assessment with survey technologies have an impact on learning gains, concluding that teacher feedback positively affects learning gains in pairs of questions. of the treatment conditions compared to the control condition. The researchers mentioned in [24] investigated the effectiveness of the application of a feedback model in the perception of teachers and students, obtaining that the feedback model of Hattie and Timperley (2007) improved the academic performance of students. Also [25] states that feedback in evaluation processes improves the quality of learning, the study shows that all students improved their performance; the training applied to teachers on how to apply feedback in online learning, improved the planning of their learning session, the learning objectives and the use of various strategies aimed at achieving the desired learning result. In reference [26] it is concluded that feedback in health sciences is a fundamental strategy in clinical learning, based on relationships, emotions and reflections, since the student faces challenging situations at the time of practice, for which affirms that the incorporation of quality feedback, aligned with the teaching-learning process, becomes important during the clinical training of the health professional. The analysis carried out by [27] in an investigation, verifies the importance of feedback considering it as a dialogical and sustainable act, they also propose peer feedback that allows self-regulation for timely learning and reference [28], concludes that in the Universities still need to improve evaluations and consider that students do not give due importance to feedback. Feedback in an online assessment system designed from a formative approach to promote and improve the learning of university students is important, because it allows identifying weaknesses in the learning process and providing timely feedback [29], in addition to its application in the teaching-learning process, it provides information that helps the university student to reduce the distance between their current learning outcome and the desired learning [30].

After analyzing the results obtained, the researcher [31] observed three feedback behaviors of the students, such as: recognizing, using and seeking feedback, originating: the beliefs, their attitudes and the perceptions of the students, including the attributescough of the teacher; the way to make the feedback; and the culture of learning. Feedback literacy, especially in health sciences students, to prepare them for a professional career, requires lifelong learning and critical thinking [32, 33].

The development of learning sessions in a virtual way, has allowed to interact with students and teachers in a synchronous and asynchronous way, the study of the investigated topic is of great importance, since its results allow corrective measures to be taken in the application of virtual feedback at the right time, while the student has the opportunity to improve their learning strategies to achieve the desired achievement. Therefore, an investigation on the feedback and the result of online learning will help corrective measures in time to provide quality teaching in coherence with the current times we live in, so the objective of the investigation was to determine the effect of the feedback in the online learning outcome of university students of health sciences.

2 Material and Method

In a non-experimental methodology, the study population consisted of 2910 students and the sample by 294, of which 225 women and 69 men, the type of intentional sampling was applied for data collection the survey technique and as instruments the questionnaire: Metacognitive Skills Inventory (MAI) designed by Shaw and Denninson in 1994. And a form for collecting information on learning outcomes (grade point average). The MAI allows students to identify metacognitive strategies (self-regulation) through of 52 items in 2 categories: 1-Knowledge of cognition 17 questions (declarative knowledge 8 questions; procedural knowledge 4 questions and conditional knowledge 5 questions) and 2-Regulation of cognition 35 questions (planning 7 questions; organization 10 questions; monitoring 7 questions, debugging 5 questions and evaluation 6 questions Prior to the application of the survey they study they gave their informed consent.

In order to process the frequencies of use of the skills, the MAI response scales were weighted, according to Table 1.

3 For data processing, a descriptive level statistical analysis was applied through the use of STUDENt's T and a regression analysis considering the regression coefficient and the standard deviation.

4 Results

The highest percentage of responses in the respondents coincides for both MAI categories on the "Agree" scale: Knowledge of cognition (61.40%) and Regulation of cognition (61.25%) (Fig. 1).

Table 1. Weighting of the scales.

Scale	Weighing
Completely agree	2
Agree	1
Neither disagree, nor agree	0
In disagreement	−1
Completely disagree	−2

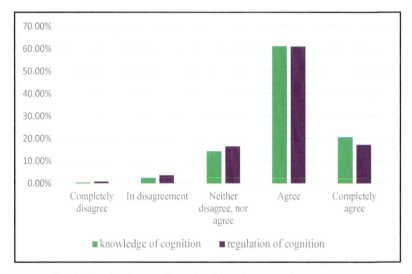

Fig. 1. Results about self-regulated learning according to category.

The frequency of use of procedural knowledge skills (65.39%) stands out slightly from the other seven subcategories of self-regulated learning in Health Sciences careers (Fig. 2).

It is noteworthy that approximately 60% of responses focus on the "Agreement" scale for subcategories of self-regulated learning (Fig. 3).

The female gender shows a higher frequency of use of skills for all the subcategories of online self-regulated knowledge than the male students surveyed, with conditional knowledge slightly standing out above the others (Fig. 4).

In the formation of professional careers in the area of Health Sciences, it seems that the female gender shows a higher learning result than the male gender.

The quartiles of the averages of the ladies are greater than the quartiles of the men; however, since the bloxpots coincide in the range of both genders, it cannot be affirmed that the female learning averages are higher (Fig. 5).

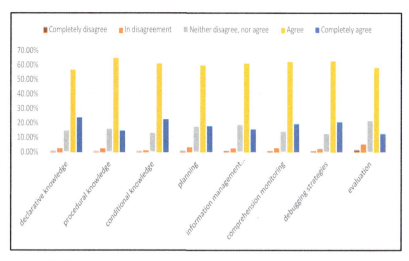

Fig. 2. Frequency of response scales by self-regulated learning subcategory.

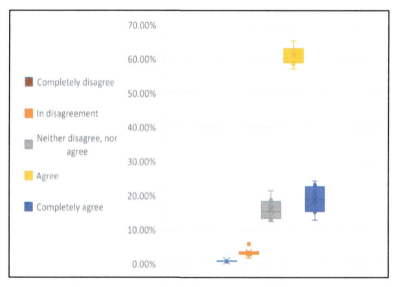

Fig. 3. Comparison between MAI subcategories and response scales according to frequency

In the regression analysis it is observed that when comparing the values obtained, the standard deviation is large in relation to the regression coefficient; therefore, the correlation is not significant (Table 2).

5 Discussion and Conclusions

When applying the inventory of metacognitive skills (MAI), to university students of Health Sciences, as shown in Figure No. 1, a result of knowledge of cognition of 61.40%

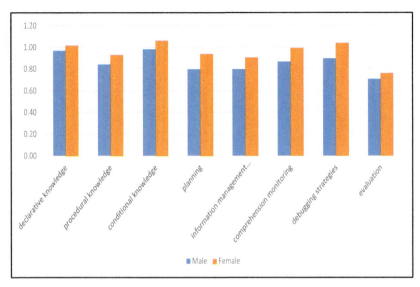

Fig. 4. Average rating of scales, according to gender and subcategories of self-regulated learning.

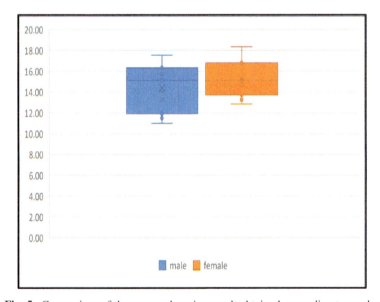

Fig. 5. Comparison of the average learning result obtained, according to gender.

and regulation of cognition 61.25 was obtained. %, obtaining only a difference of 0.15%, when compared with the feedback study carried out by [22] in online learning, it is observed that automatic feedback increases student performance by 65% and automatic feedback is more efficient in relation to the manual in 82%. Investigated [23] and concluded that teacher feedback positively influences learning gains in paired questions

Table 2. Relationship between self-regulated learning skills and rating scales.

		Completely disagree	In disagreement	Neither disagree, nor agree	Agree	Completely agree
Knowledge about cognition	Coef. Desv. Est	−1.1 (±2.5068)	−0.2 (±0.9592)	−0.3 (1.1331)	0.9	1.6 (±1.1196)
Regulation of cognition	Coef. Desv. Est	3.9 (±3.6536)	−1.2 (±2.467)	0.7 (±1.4433)	−0.5	−1.9 (±1.4387)

compared to the control-only condition. Like this research [8], they used a feedback training strategy, developed by the technical team of the Teaching Improvement Unit (UMD) for the students. Obtaining the following results: University Selection Test (PSU) was not related to academic performance. It was detected that motivational learning strategies, anxiety management and study planning should be strengthened in students, and in relation to cognitive strategies, precision and accuracy, systematic exploration and linguistic skills, they were also considered. Important.

In Fig. 2, we observe that procedural knowledge skills reached 65.39%, rising slightly from the other 7 categories. In the research carried out by [17], they reviewed the general principles of teaching in clinical settings, such as identifying the needs of students, teaching according to a model and then feedback. By using systematic methodologies with evidence of validity, for teaching in clinical contexts, it allowed a more complete learning and feedback. Researchers [11, 13] report that students commented that they had difficulty observing and understanding the live demonstrations, however their comments were positive regarding the demonstration videos, when used in the classroom and later when posted online. Through the videos, the performance of the students rose. According to [12] when providing the video postcasts to the students, there was an increase in their performance of the students, they stated that they were very useful, improving their skills and facilitating their learning. In [34] it refers that feedback is perceived as a teaching-learning process in clinical laboratories; students perceived the feedback difference in clinical settings; Through interpersonal interrelationships, the nursing students valued the feedback received in the clinical laboratories, which was group, and also the emotional reactions that originated. They concluded that students were more open to feedback in clinical settings.

In the research work presented in Fig. 4, there is evidence of a slight increase in the female gender in conditioned knowledge, just as the author [6] considers gamification one of the most innovative teaching strategies today, being the purpose of the investigation: to increase the motivation of the students and to awaken their interest towards the subjects. In the subjects of Health Sciences, real clinical practice is very important, and that the teacher presents the real clinical cases, and grouping the students into small teams, and thus collaboratively solve the cases When using the gamification strategy rapid and continuous feedback is necessary to increase the participatory motivation of students.

Other authors such as [7] consider motivation as a strategy for good results, suggesting an alternative in the teaching-learning process for the area of health sciences: the guided university debate.

According to the author [31], feedback literacy skills in students should be improved, it is considered one of the keys, also to train them and they can recognize, seek and use the feedback provided by the university teacher. Likewise [32, 33] makes us reflect on the importance of feedback, which allows improving academic performance, its provision by teachers is often not enough by itself. The results were: the commitment of the students with the first phase of the study was poor, in the second phase they provided the opportunity to participate in the development of a set of modules, which were designed to encourage teachers and students in collaborative work, to incentivize skills, all teams reported short-term and long-term benefits with this approach.

It is concluded that feedback in university students of Health Sciences in virtual environments is effective when applied in a timely manner and can be planned, based on the evidence of the learning outcome.

To achieve this, they must be previously trained, from the first semesters of study, in feedback literacy, making it part of the self-regulation of their learning.

References

1. Canabal, C., Margalef, L.: La retroalimentación: la clave para una evaluación orientada al aprendizaje. Profesorada Revista de curriculum y formación del profesorado **21**(2), 149–170 (2017)
2. Min, Q., Chen, Y., Liu, N., Zuo, M.: A learning style model designed for online learning environments. Int. J. Inf. Educ. Technol. **8**(9), 623–627 (2018)
3. Huertas Bustos, A.P., Vesga Bravo, G.J., Galindo León, M.: Validación del instrumento Inventario de habilidades metacognitivas (mai)'con estudiantes colombianos. Praxis Saber **5**(10), 56–74 (2014)
4. Cui, Y.: Self-efficacy for self-regulated learning and Chinese students' intention to use online learning in COVID-19: a moderated mediation model. Int. J. Inf. Educ. Technol. **11**(11), 532–537 (2021)
5. Ramírez, M.D.L.L.B., Hernández, L.G.J.: Diseño y validación de un instrumento para evaluar la retroalimentación asertiva en educación normal. IE Revista de Investigación Educativa de la REDIECH **11**, 791 (2020)
6. Sampedro Piquero, P.: Gamificación en el aula universitaria: La liga de los casos clínicos en Psicología (2020). https://riuma.uma.es/xmlui/handle/10630/19572
7. Arrue, M., Zarandona, J.: El debate en el aula universitaria: Construyendo alternativas para desarrollar competencias en estudiantes de ciencias de la salud. Educación Médica **22**, 428–432 (2021). https://doi.org/10.1016/j.edumed.2019.10.016
8. Garces Bustamante, J., Labra Godoy, P., Vega Guerrero, L.: La retroalimentación: una estrategia reflexiva sobre el proceso de aprendizaje en carreras renovadas de educación superior. Cuadernos de Investigación Educativa **11**(1), 37–59 (2020). https://doi.org/10.18861/cied.2020.11.1.2942
9. Iglesias Martínez, M.J., Lozano Cabezas, I., Martínez Ruiz, M.Á.: La utilización de herramientas digitales en el desarrollo del aprendizaje colaborativo: análisis de una experiencia en Educación Superior. Red U: revista de docencia universitaria (2013)
10. Cabero-Almenara, J., Palacios-Rodríguez, A.: La evaluación de la educación virtual: Las e-actividades. RIED. Revista Iberoamericana de Educación a Distancia **24**(2), 169–188 (2021)

11. Rose, T.M.: Lessons learned using a demonstration in a large classroom of pharmacy students. Am. J. Pharm. Educ. **82**(9), 6413 (2018). https://doi.org/10.5688/ajpe6413.PMID:30559495; PMCID:PMC6291665
12. Mnatzaganian, C.L., Singh, R.F., Best, B.M., Morello, C.M.: Effectiveness of providing video podcasts to pharmacy students in a self-study pharmaceutical calculations module. Am. J. Pharm. Educ. **84**(12), ajpe7977 (2020). https://doi.org/10.5688/ajpe7977
13. Suartama, I.K., Mahadewi, L.P.P., Divayana, D.G.H., Yunus, M.: ICARE approach for designing online learning module based on LMS. Int. J. Inf. Educ. Technol. **12**(4), 305–312 (2022)
14. Chansanam, W., Tuamsuk, K., Poonpon, K., Ngootip, T.: Development of online learning platform for Thai University students. Int. J. Inf. Educ. Technol. **11**(8), 348–355 (2021)
15. Franco-Coffré, J.A., Mena-Martin, F., Gordillo-Ojeda, M.V., Vargas-Aguilar, G.M.: La Educación virtual en la formación profesional de enfermeros, durante la pandemia provocada por la COVID 19. Polo del Conocimiento **6**(12), 762–775 (2021)
16. Yusef Contreras, V.A., Sanhueza Ríos, G.A., Seguel Palma, F.A.: Importancia de la simulación clínica en el desarrollo personal y desempeño del estudiante de enfermería. Ciencia y enfermería **27** (2021)
17. Gutiérrez-Cirlos, C., Naveja, J.J., Sánchez-Mendiola, M., Gutiérrez-Cirlos, C., Naveja, J.J., Sánchez-Mendiola, M.: Modelos de educación médica en escenarios clínicos. Investigación en educación médica **9**(35), 96–105 (2020). https://doi.org/10.22201/facmed.20075057e.2020.35.20248
18. Després-Bedward, A., Avery, T.L., Phirangee, K.: Student perspectives on the role of the instructor in face-to-face and online learning. Int. J. Inf. Educ. Technol. **8**(10), 706–712 (2018)
19. Rodríguez, S., Condés, E., Arriaga, A.: Irrupción de la simulación clínica online en tiempos de COVID-19. Una experiencia ilustrativa de asignatura en el Grado de Psicología. FEM: *Revista de la Fundación Educación Médica* **24**(2), 101–104 (2021)
20. Zuluaga-Gómez, M., Valencia-Ortiz, N.L.: Educación en facultades de medicina del mundo durante el periodo de contingencia por SARS-COV-2. Med UNAB **24**(1), 92–99 (2021)
21. González-Fernández, D., Gambetta-Tessini, K.: Estrategias para potenciar la retroalimentación en los talleres disciplinares de las carreras de Ciencias de la Salud. Educación Médica **22**, 283–287 (2021)
22. Cavalcanti, A.P., et al.: Automatic feedback in online learning environments: a systematic literature review. Comput. Educ. Artif. Intell. **2**, 100027 (2021). https://doi.org/10.1016/J.CAEAI.2021.100027
23. Molin, F., Haelermans, C., Cabus, S., Groot, W.: Do feedback strategies improve students' learning gain?—Results of a randomized experiment using polling technology in physics classrooms. Comput. Educ. **175**, 104339 (2021). https://doi.org/10.1016/J.COMPEDU.2021.104339
24. Cano, L.G.: Percepción de profesores y estudiantes universitarios sobre la retroalimentación y su incidencia en el rendimiento académico a partir del uso de un modelo de retroalimentación (2017)
25. Valenzuela-Valenzuela, D., Bastías-Vega, N., Pérez-Villalobos, C.: Resultados de una capacitación sobre retroalimentación efectiva para tutores clínicos de internado de universidades chilenas. FEM: *Revista de la Fundación Educación Médica* **24**(4), 183–190 (2021)
26. Fuentes Cimma, J., Ortega, J., Parra, P., Villagrán, I., Isbej, L., Leiva, I.: Feedback: Un pilar fundamental del aprendizaje clínico: Feedback: A cornerstone of clinical learning. ARS MEDICA *Revista De Ciencias Médicas* (2021). https://doi.org/10.11565/arsmed.v46i4.1851
27. Quezada Cáceres, S., Salinas Tapia, C.: Modelo de retroalimentación para el aprendizaje: Una propuesta basada en la revisión de literatura. Rev. Mex. Investig. Educ. **26**(88), 225–251 (2021)

28. Báez-Rojas, C., Córdova-León, K., Fernández-Huerta, L., Villagra-Astudillo, R., Aravena-Canese, L.: Modelo de retroalimentación mediante evaluación de 360° para la docencia de pregrado en ciencias de la salud. FEM: *Revista de la Fundación Educación Médica* **24**(4), 173–181 (2021)
29. Navaridas-Nalda, F., González-Marcos, A., Alba-Elías, F.: Evaluación online orientada al aprendizaje universitario: Impacto del feedback en los resultados de los estudiantes. Revista Interuniversitaria de Formación del Profesorado **34**(2), 101–120 (2020)
30. Prada, M.F.H., Zamudio, C.D.P.A., Pacherres, Y.C.V., Bruno, C.R.B.: La retroalimentación formativa una práctica eficaz en tiempos de pandemia. Horizontes. *Revista de Investigación en Ciencias de la Educación* **5**(21), 1480–1490 (2021)
31. Bowen, L., Marshall, M., Murdoch-Eaton, D.: Medical student perceptions of feedback and feedback behaviors within the context of the "Educational Alliance." Acad. Med. J. Assoc. Am. Med. Colleges **92**(9), 1303–1312 (2017). https://doi.org/10.1097/ACM.0000000000001632
32. O'Connor, A., McCurtin, A.: A feedback journey: employing a constructivist approach to the development of feedback literacy among health professional learners. BMC Med Educ. **21**(1), 486 (2021). https://doi.org/10.1186/s12909-021-02914-2.PMID:34503487;PMCID: PMC8429041
33. Ulfatun, T., Septiyanti, F., Lesmana, A.G.: University students' online learning self-efficacy and self-regulated learning during the COVID-19 pandemic. Int. J. Inf. Educ. Technol. **11**(12), 597–602 (2021)
34. Nuuyoma, V.: Feedback in clinical settings: nursing students' perceptions at the district hospital in the southern part of Namibia. Curationis **44**(1), e1–e12 (2021). https://doi.org/10.4102/curationis.v44i1.2147

Educational Management, Psychology and Educational Statistics

Exploration and Application of the Blended Learning Model in the "Software Engineering" Course

Mengmei Wang(✉)

School of Artificial Intelligence and Software, Jiangsu Normal University Kewen College,
Xuzhou City 221132, China
mengmei_wang@cumt.edu.cn

Abstract. Software Engineering is a brand-new engineering course in the application-oriented skills training program, and an essential professional skill for high-demand employment. Because of the fast growth of science and technology, many students have high expectations for this topic. Students' excitement and expectations for this topic, however, are dwindling as a result of uninspiring texts and an antiquated teaching method. To improve the problem, this study innovates in the areas of teaching idea, teaching substance, teaching style, and assessment mode. Based on the education concept of "student-centered" and "project-driven," three basic teaching modalities of "role-playing," "project case teaching," and "workshop" were chosen to optimize the teaching content and assessment procedures. The blended teaching method shifts the function of learning from passive to active, and students are liberated from the monotonous theoretical learning environment. Students have made significant progress in course scores, competitions, projects, and other areas as a result of practical instruction, and the effect is palpable.

Keywords: blended learning model · Software engineering · Workshop

1 Problem Existing in the Teaching of Software Engineering

1.1 The Educational Topic is not Properly Integrated with Industry, and There is a Lack of Training for Engineering Practice

During the course of their education, kids are spoon-fed passively. This learning technique does not successfully participate in learning. Critical thinking, teamwork, and creative thinking opportunities are being lost. The primary mode of instruction is the classroom, and real-world engineering practice, interpersonal contact, and social activities are insufficient. The approach for assessing students' learning effects is pretty easy. The theoretical knowledge test is still used as a final evaluation technique. Process evaluation, such as practical ability assessment, is lacking. Students' general engineering practice ability [1–3] needs to be enhanced as an engineering major. Some courses and textbooks' material is insufficiently integrated with industry, or even disjointed, necessitating the introduction of more practical teaching techniques [4–6].

1.2 Teaching Output is Out of Sync with Societal Requirements and Disconnected from a Continuously Changing World

The product of education is for societal purposes, but presently the training of teaching skills is disconnected from societal demands.

1.3 The Utilization Rate of Teaching Resources is Low, and the Teaching Means in Traditional Classroom Teaching Are Limited and Rigid

Conventional teaching stresses the role of instructors as leaders. The form is very basic and structured, neglecting instructional channels outside the classroom and instructing in line with their aptitude. Learning websites such as Xuetang Online [7, 8], Chaoxing Fanya [9, 10], Yuclassroom [11], I wish to Teach myself, and Bilibili [12–15] have proliferated, as have a huge number of online education teachers (specializing in online training) [16]. However, the majority of instructors at the school continue to follow the guidelines and teach in the traditional classroom setting. As a result of this phenomenon, a rising number of students are unwilling to listen to lectures and instead rely on internet resources for self-study. This is hardly a surprise to teachers or schools.

2 In a Software Engineering Course, Blended Teaching is Used

2.1 It Stresses the "STudent-Centered" and "Project-Driven" Teaching Concepts in Engineering Education

To boost students' ability to practice engineering [17], "practice" should be elevated to the top of the priority list. The course teaches various system products as project subjects, using actual project development as the primary teaching method and evaluation tool. It not only increases student interest, but also secures the project's practicality and long-term viability. The "workshop" model [18] was also introduced at the same time. To complete the entire process of "searching for problems, assessing problems, offering solutions, achieving solutions, and presenting solutions," the method requires participation in the time allotted. Furthermore, in this mode, every team member participates and is scored based on their performance. As a result, team members can completely collaborate while yet competing in the background. The survey is structured according to the following principles: "social survey—determination of main job positions—analysis of typical tasks—analysis of competence fields—construction of learning fields" [19–21].

2.2 Reconstruction and Optimization of Curriculum Content Based on "Social Needs" + "Student Needs"

The course content is combined with the requirements for social media posts. The curriculum system is equivalent to the vocational ability system, and the training process is linked to the work process. Fully reflect the requirements of the software engineering profession as well as social development [22–25]. We conducted a significant amount of social research. The main jobs of software engineering and technology specialization, as well as their typical tasks, are carefully analyzed and studied.

Students should be given individualized attention following the educational objectives [26, 27]. As a result, it may assist students in defining the knowledge structure, quality structure, and ability structure that they should have for future growth. Students create their own four-year tailored learning and growth plan. It may help and lead pupils to grow up autonomously, develop separately, and realize their uniqueness based on generality. Students' successes in "discipline competition" [28, 29], "cultural and sports activities", "social work", "innovation and entrepreneurship" [30], "skills examination" and other areas can be traded for equivalent individualized development credits. Students might be encouraged to achieve their development in this manner.

2.3 Increase the Exploitation of Teaching Materials and Develop a Smart Teaching Method

Rain classroom and superstar are used to complement one other in-classroom education to make apparent and fulfill the teaching purpose. First and foremost, the main classroom is utilized for attendance, questionnaire surveys, and exercise consolidation. If any students like to consolidate the teaching information after the teaching session, superstar software will be utilized to create online video learning. Students' online learning and classroom discussions may be dynamically followed via online video learning. The final test will be conducted without the use of paper. Superstar is used to creating the exam question library for the software engineering course. Students can come in at any time to learn something new while also becoming acquainted with the final test outline and approach.

2.4 Big Data Analysis of Students' Learning Situations, Gap Identification, and Filling, and Feedback Teaching

Educational evaluation will transition from a "results view" to a "process perspective" with the use of big data technologies. Evaluate students' learning traits, strengths, and potential via supplemental data collecting and automated data analysis. Provide learning analysis reports to each student to suit the development demands of pupils with varying abilities. It can assist them in continually improving themselves.

3 Arrangement of the Teaching Process in the Mixed Teaching Style

This article uses the "life cycle of software development" section as an example to construct the scenario simulation teaching method and heuristic teaching mode to discuss the implementation of hybrid teaching mode in practical courses in detail. "Problem introduction—scenario recurrence (workshop)—problem summary" is the teaching procedure.

First and foremost, from the elaboration of software challenges to the inspiration of software engineering concepts. Case analysis and explanation can assist students in consolidating their understanding of each level. Second, the resolution of the software dilemma leads to the section's primary material. The "workshop" mode methodically displays the six steps of software development planning, requirements analysis, design, coding, testing, and maintenance. This hybrid technique can boost students' enthusiasm for learning while simultaneously improving their practical cognition of the educational subject. On the one hand, it allows students to enjoy the benefits of collaboration while still developing their abilities to investigate and solve challenges individually. On the other hand, it builds a solid basis for children to freely investigate, think about, and solve problems in the future. Furthermore, by combining multimedia and conventional teaching methods, students' interest, initiative, and creativity in learning may be increased. So that kids may actively participate in learning in several ways. It can increase students' enthusiasm for studying while also gradually increasing the difficulty of the problem to reach the objective of boosting students' scientific excellence.

Bridge-in, objective, pre-test, participation, post-test, and summary are the several stages of the teaching process.

3.1 Bridge-In

Teaching Arrangement. The teacher began the new class by showing the film "People's Software Understanding."

Teachers' Activities. Activity 1: After showing the film "People's Understanding of Software," the instructor poses the heuristic question "What is software?" Finally, the video's software is summarized.

Student Activities. Activity 1: Students view video snippets and consider the notion of software.

Design Concept. The design notion begins with the reality around us and progresses to an abstract concept.

3.2 Objective

Teaching Arrangement. The teacher explains the curriculum objectives and theoretical material, as well as major elements of ideological and political thought.

Teachers' Activities. Activity 1: The teacher distributed the "Rain classroom" voting job and outlined students' issues with the program. Activity 2: Case Analysis 1. The teacher, ARIANE5 Rocket, provides the case facts and summarizes the rationale. Activity 3: Case Analysis 2, The teacher provides the case facts and summarizes the rationale using the Windows Vista system. Activity 4: Case analysis 3, The teacher provides the case material and discusses the reasons for the 12306 network ticket purchase system. Activity 4: Teachers can present the idea, primary manifestations, and causes of the software crisis through these scenarios.

Student Activities. Activity 1: Students take the initiative to vote in "Rain Class" and express their viewpoints. Activity 2: Students consider the commonality of examples, the reasons, and solutions under the supervision of case-based education. Activity 3: Students have a basic comprehension of the idea, as well as the primary forms and causes of the software crisis. Activity 4: How can the software crisis be resolved?

Design Concept. This can help kids improve their fundamental reading skills.

3.3 Pre-test

Teaching Arrangement. First and foremost, in response to the heuristic question "how to fix the software crisis?" This course introduces the notion of software engineering. Second, a software engineering solution for resolving the software issue is proposed. Finally, the software development life cycle idea and its six stages are introduced.

Teachers' Activities. Activity 1: The heuristic question "How to address the software crisis?" was posed by the teacher. As a result, the idea of software engineering is presented. Activity 2: The teacher shows the movie "People's perspectives on software engineering" and outlines the significance and need for software engineering. Activity 3: The notion of the software development life cycle is presented by the teacher. The standpoint of dialectical materialism, also defines the link between the software development life cycle and the software crises. Activity 4: Teachers used role-playing to model real-world development problems. Six stages of the software development life cycle are discussed in depth.

Student Activities. Activity 1: Pay attention to the teacher and consider the topic of "how to fix the software crisis?". Activity 2: Students see the film and realize the significance of software engineering. Activity 3: The heuristic question "How does software engineering tackle the software crisis?" is considered by students. Activity 4: Students learn about the software development life cycle. Activity 5: Students observe role-playing development scenarios and gain an understanding of the whole software development life cycle.

Design Concept. Under the direction of dialectical materialism theory, the instructor employs the approach of heuristic inquiry to teach, gradually engaging students' learning attention. Teachers use role-playing to replicate genuine developing scenarios under the direction of new ideas to boost students' autonomous learning capacity for discovering information.

3.4 Participation

Teaching Arrangement. The "workshop" teaching approach is used to provide students with a more in-depth understanding of the software development life cycle.

Teachers' Activities. Activity 1: The teacher assigns the "Workshop" topic of handheld campus APP creation to the students. Activity 2: The workshop is taught by the teacher in a teacher-led format.

Student Activities. Activity 1: Students work in groups to discuss their topic. Students develop and discuss in phases in Activity 2.

Design Concept. Students can get an understanding of the software development life cycle and boost their interest in this subject by using the workshop format.

3.5 Post-test

Teaching Arrangement. Student representatives present the workshop's outcomes.

Teachers' Activities. Activity 1: The teacher presides over the process of sharing student achievements.

Student Activities. Activity 1: Students present their accomplishments in groups.

Design Concept. Students may acquire a sense of recognition and accomplishment in learning by sharing their achievements. This allows them to expand their grasp of the software development cycle, exercise their expressive ability, and obtain a sense of recognition and achievement in their learning.

3.6 Summary

Teaching Arrangement. In this session, the teacher summarizes and extends the course by displaying a PPT of knowledge points.
Teachers' Activities. Activity 1: The instructor organizes everything he has learned in class and makes a summary statement.
Student Activities. Activity 1: Students listen to and review the lesson material. Activity 2: Students consider software engineering application scenarios.
Design Concept. This assists students in clarifying the knowledge system and achieving comprehensive knowledge meaning creation.

4 Teaching Outcomes in the Mixed Modality of Instruction

The complete and systematic advances in the four elements of an idea, educational content, educational model, and teaching evaluation are the teaching innovation accomplishments of the software engineering course. The implementation of successes has yielded tremendous outcomes during the last two years. This article will go through the implementation of curriculum teaching innovation as well as the ripple impact of curriculum reform.

4.1 The Using Course Teaching Innovation

The evaluation of pupils' learning state is more thorough. This course lays the groundwork for the Rain-classroom and the superstar dual assessment technique. When online courses are delivered, students are extensively assessed through online learning on the Superstar platform, the release of tasks, and the completion of unit examinations. Teachers teach the curriculum in person during offline courses, while students are assessed by attendance tracking, sharing courseware, class roll call, and real-time assignment of exam questions in rain class. After years of complimenting each other, it has been able to master students' learning situations in real-time, and change the difficulty and substance of teaching based on students' unique learning situations, resulting in the teaching impact of individualized and comprehensive teaching in Fig. 1.

Students' practical skills have been honed. The general growth in students' professional skills is evident, particularly in discipline contests in Fig. 2. It won a total of 26 national and provincial competition honors between 2020 and 2021, including 13 national and 13 provincial awards. A total of 34 students received prizes, with 91% of the winners taking software engineering courses, which directly and indirectly made all types of entries more market-oriented, mature, and standardized.

The students have made outstanding contributions to scientific research and innovation. Students have successfully applied for 19 provincial innovation projects

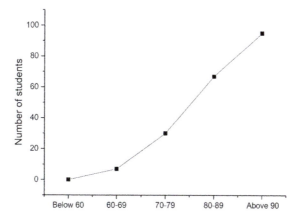

Fig. 1. Analysis chart of final exam results

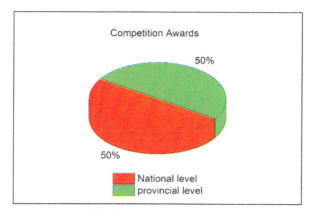

Fig. 2. Competition awards

and 30 college-level innovation projects in and out of software engineering classroom teaching during the last two years in Fig. 4. A total of 71 students sponsored or participated in provincial innovation initiatives, with 65 majoring in software engineering making up 91.54% of the total in Fig. 3. So far, I have assigned three journal publications to students, four research reports to undergraduates, and two software copyrights to students.

4.2 Curriculum Reform's Radiant Effect

Curriculum reform's associated outcomes are beginning to emerge. IT overcomes the problem of software engineering professionals' ability training not matching the criteria of IT companies, and it completes the training requirements of "the final mile" from the school to the company. It achieves direct docking with the growth of the software industry and the need of the information industry thanks to its novel qualities of "complete

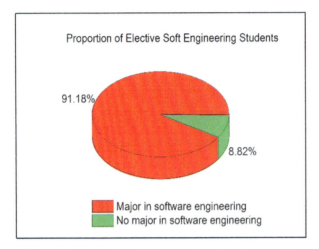

Fig. 3. The proportion of elective soft engineering students

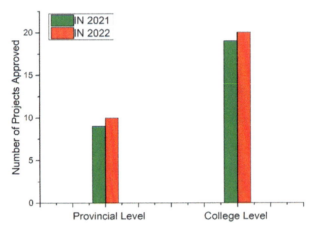

Fig. 4. Province-level and college-level projects

support of the software industry and deep integration of the information industry." Construction of the next generation of information technology specialist group integrated with software engineering specialty construction, depending on Neusoft to carry out in-depth integration of industry and education. The "multi-certificate" and "competition to promote education and learning" systems have enhanced education quality, and an incubator platform and mechanism for innovation and entrepreneurial education have been built.

The course instruction includes a demonstration and radiation impact. Years of effort have gone into the teaching reform of the software engineering course, and certain experience and methods have been accumulated, from which students not only gain design skills and theories, but also experience the current working atmosphere, pressure, and cooperation rules in the software company in advance, making students more accurate in their positioning and clearer in the direction of job hunting.

5 Conclusion

Is prioritized with "workshop," "role play," and "project" case teaching auxiliary hybrid teaching mode will learn the role from passive to active, students learn from the theory of boring environment, in-depth experience in software engineering of the project development process of the analysis, design, coding, testing, and software project management, software quality management, The engineering concept of software configuration management improves students' theoretical and practice application ability, collaborative communication ability, issue analysis, and solution ability, and so engineering practice ability.

References

1. Zeng, J., Liu, L., Tong, X., et al.: Application of blended teaching model based on SPOC and TBL in dermatology and venereology. BMC Med. Educ. **21**(1), 1–7 (2021)
2. Şentürk, C.: Effects of the blended learning model on preservice teachers' academic achievements and twenty-first century skills. Educ. Inf. Technol. **26**(1), 35–48 (2021)
3. Archibald, D.E., Graham, C.R., Larsen, R.: Validating a blended teaching readiness instrument for primary/secondary preservice teachers. Br. J. Edu. Technol. **52**(2), 536–551 (2021)
4. Islam, M.K., Sarker, M.F.H., Islam, M.S.: Promoting student-centred blended learning in higher education: a model. E-Learn. Digit. Media **19**(1), 36–54 (2022)
5. Jalinus, N.: Developing blended learning model in vocational education based on 21st century integrated learning and industrial revolution 4.0. Turkish J. Comput. Math. Educ. (TURCOMAT), **12**(8), 1239–1254 (2021)
6. Dai, N.V., Trung, V.Q., Tiem, C.V., et al.: Project-based teaching in organic chemistry through blended learning model to develop self study capacity of high school students in Vietnam. Educ. Sci. **11**(7), 346 (2021)
7. Giray, G.: An assessment of student satisfaction with e-learning: an empirical study with computer and software engineering undergraduate students in Turkey under pandemic conditions. Educ. Inf. Technol. **26**(6), 6651–6673 (2021)
8. Mielikäinen, M.: Towards blended learning: Stakeholders' perspectives on a project-based integrated curriculum in ICT engineering education. Ind. Higher Educ. 0950422221994471 (2021)
9. Ye, L., Zhong, J.: Study on blended teaching in principles of chemical engineering based on cloud platform. In: IOP Conference Series: Earth and Environmental Science. IOP Publishing, vol. 693, no. 1, pp. 012027 (2021)
10. Sasmito, A.P., Kustono, D., Purnomo, P., et al.: Development of android-based teaching material in software engineering subjects for informatics engineering students. Int. J. Eng. Pedagog. **11**(2), 25–40 (2021)
11. Fitoussi, R., Chassidim, H.: Teaching software engineering during covid-19 constraint or opportunity. In: 2021 IEEE Global Engineering Education Conference (EDUCON). IEEE, pp. 1727–1731 (2021)
12. Ståhl, D., Sandahl, K., Buffoni, L.: An eco-system approach to project-based learning in software engineering education. IEEE Trans. Educ. (2022)

13. Nash, J.A., Pritchard, B.P.: Coding, software engineering, and molecular science−teaching a multidisciplinary course to chemistry graduate students. In: Teaching Programming Across the Chemistry Curriculum. American Chemical Society, pp. 159–171 (2021)
14. Adil, M., Fronza, I., Pahl, C.: Software design and modeling practices in an online software engineering course: learners' perspective (2022)
15. Tian, L., Yang, Y., Han, Z., et al.: Teaching design and practice of software engineering series courses based on OBE. In: 2021 2nd International Conference on Computers, Information Processing and Advanced Education, pp. 681–686 (2021)
16. Sharafuddin, H., Allani, C.: Evaluation of the blended learning system in higher education: AOU – Kuwait. Int. J. Inf. Educ. Technol. **2**(4), 412–414 (2012)
17. Chaiyama, N.: The development of blended leaning model by using active learning activity to develop learning skills in 21st century. Int. J. Inf. Educ. Technol. **9**(12), 880–886 (2019)
18. Juhary, J.: Perceptions of students: blended learning for IR4.0. Int. J. Inf. Educ. Technol. **9**(12), 887–892 (2019)
19. He, W., Zhao, L.: Exploring undergraduates' learning engagement via BYOD in the blended learning classroom (EULEBYODBLC). Int. J. Inf. Educ. Technol. **10**(2), 159–164 (2020)
20. Li, N., Wang, J., Zhang, X., Sherwood, R.: Investigation of face-to-face class attendance, virtual learning engagement and academic performance in a blended learning environment. Int. J. Inf. Educ. Technol. **11**(3), 112–118 (2021)
21. Chauhan, S., Naseem, A., Rashwan, E.: Developing a quality checklist for designing blended learning course content. Int. J. Inf. Educ. Technol. **6**(3), 224–227 (2016)
22. Tongchai, N.: Impact of self-regulation and open learner model on learning achievement in blended learning environment. Int. J. Inf. Educ. Technol. **6**(5), 343–347 (2016)
23. Min, Q., Guanghui, W.: A blended learning strategy for professional English course in a cloud learning environment. Int. J. Inf. Educ. Technol. **7**(8), 608–611 (2017)
24. Zhang, W., Zhu, C.: Review on blended learning: identifying the key themes and categories. Int. J. Inf. Educ. Technol. **7**(9), 673–678 (2017)
25. Shu, J., Hu, Q., Zhi, M.: Research on the learning behavior of university students in blended teaching. Int. J. Inf. Educ. Technol **9**(2), 92–98 (2019)
26. Lawn, M.J., Lawn, E.: Increasing English communicative competence through online English conversation blended e-learning. Int. J. Inf. Educ. Technol. **5**(2), 105–112 (2015)
27. Rodmunkong, T.: The development of blended learning using internet in computer programming and algorithm. Int. J. Inf. Educ. Technol. **5**(6), 442–446 (2015)
28. Chaiyama, N.: The development of blended learning management model in developing information literacy skills (BL-ILS model). Int. J. Inf. Educ. Technol. **5**(7), 483–489 (2015)
29. Au Thien Wan: How can learners learn from experience? a case study in blended learning at higher education. Int. J. Inf. Educ. Technol. **5**(8), 615–619 (2015)
30. Yeen-Ju, H.T., Mai, N., Selvaretnam, B.: Enhancing problem-solving skills in an authentic blended learning environment: a Malaysian context. Int. J. Inf. Educ. Technol. **5**(11), 841–846 (2015)

Improving Learning Outcomes with Pair Teaching StrateFiggy in Higher Education: A Case Study in C Programming Language

Yongbin Zhang[1], Ronghua Liang[1], Yuansheng Qi[1], Xiuli Fu[2(✉)], and Yanying Zheng[3]

[1] Beijing Institute of Graphic Communication, Beijing, China
{zhangyongbin,liangronghua,yuansheng-qi}@bigc.edu.cn
[2] Beijing Institute of Petrochemical Technology, Beijing, China
fuxiuli@bipt.edu.cn
[3] Beijing University of Agriculture, Beijing, China
huaxue@bua.edu.cn

Abstract. Learning outcomes have attracted more and more attention in higher education. Many teaching and learning methods have been invented to improve learning outcomes. Teaching and learning pedagogies will attract intensive focus because the educational paradigm is moving from teacher-centered to student-centered learning. However, there is little research on improving learning outcomes based on existing teaching and learning contexts. This paper presented a creative strategy to enhance learning outcomes with two instructors teaching by turns based on learning theories. This novel method utilized the characteristics of different instructors to facilitate student learning and was adopted in the C programming language course to verify its effectiveness. Participants were divided into two groups according to their majors. We randomly selected one group as the treatment group, and the left group was the control group. One instructor taught the control group, and another lectured the treatment group. Students took the pre-test after twelve weeks. The instructor from the control group led the treatment group for a week before the end of the course. Students who failed the pre-test took the post-test. Our experimental results showed that the pair instructing strategy could increase learning outcomes.

Keywords: C programming language · Higher education · Improving · Learning outcomes · Pair teaching · Pedagogy

1 Introduction

Learning outcomes in higher education have gained growing attention globally [1]. The outcome-based education theory has impacted higher education significantly [2]. The Organisation for Economic Co-operation and Development (OECD) has initiated the project to measure learning outcomes at the international level. For national-level assessments, the Collegiate Learning Assessment (CLA) is one of the most popular tests for measuring learning outcomes in the United States [3]. A similar test has been developed by the Australian Council for Educational Research (ACER) and is widely used in Australia [4]. In addition to these international and national tests, assessments for programs and courses also capture the attention of researchers. Researchers developed a framework for comprehensive program evaluation based on the balanced scorecard and organizational frames [5]. Another method of assessing single course learning outcomes has been designed based on the idea that each learning outcome should be evaluated by different tools [6].

Apart from evaluating learning outcomes, methods or pedagogies for enhancing learning outcomes are another highlight spot in higher education. Simulations can encourage students to engage in a productive exploration of scientific phenomena and improve learning outcomes in physics education when carefully developed and thoroughly tested [7]. Social constructivism theories and experiment learning depict how simulation-based education improves understanding [8]. Research also shows that interactive learning styles can improve learning outcomes in large classes [9]. Apart from cognitive and education science, artificial intelligence algorithms have also been adopted to enhance learning outcomes [10]. The emphasis on learning outcomes requires the shift from teacher-centered teaching to student-centered learning. But it is difficult for all teachers to carry out the new paradigm in practice [11].

To minimize the difficulty, we provide a pair teaching strategy to enhance learning outcomes in this paper. Teachers do not need extra effort to implement the pair teaching method.

2 Literature Review

2.1 Learning Outcomes

Despite the extensive focus on learning outcomes, a consistent learning outcomes definition is lacking. Learning outcomes have often been placed where the term learning objectives should be in published research. For example, learning outcomes mean what the learner should be able to do upon the end of the learning activities, while outcomes of learning describe the actual gains of learners from their studies [12]. Similarly, another author adopted learning outcomes as what students are expected to demonstrate in terms of knowledge, skills, and attitudes at the end of the learning experience [6]. In the book, a learning outcome has been defined as an expressed expectation of what someone will have learned [13].

The lack of a uniform definition prevents effective learning and teaching in higher education, particularly when learning outcomes are misunderstood as learning objectives. Although some researchers regard learning outcomes as another name for learning objectives, the two terms are different; and their differences impact teachers and students [14]. Teachers would like to focus on the course content if they mistake learning outcomes for learning objectives. With traditional teacher-centered methods, instructors decide on course content to deliver and design assessments to test how well the students learned about the course [11]. Therefore, teachers who interpret learning outcomes as learning objects would believe students achieve the learning outcomes as long as specified knowledge has been covered, which is the primary issue in teacher-centered methods. The student-centered learning paradigm is presently stepping on the state in higher education. This change means what students can do attracts more attention [11].

It is urgent to clarify the differences between learning outcomes and learning objectives. According to the Cambridge dictionary, an outcome is a result or effect of an action, situation, etc., while an objective is something planned to do or achieve. Therefore, we agree that learning outcome is what a learner achieves after learning, which has a similar meaning to the term outcomes of learning - actual achievements of students as a result of their learning activities [12]. On the other hand, a learning object is what students are expected to achieve at the end of their learning. Therefore, Learning outcomes and objectives are closely related to each other. Learning outcomes are measured or assessed after teaching and learning activities. Learning objects are designed before the activities. Differences between them are shown in Fig. 1.

Differences between two terms	
Learning Objectives	Learning Outcomes
Describe what the learner should be able to do upon completion of an educational activity	Describe the learner's actual achievements as a result of the Educational activity
Can be designed	Should be evaluated
Are intended learning outcomes	Could equal, be more or less than what expected

Fig. 1. Differences between learning objectives and outcomes.

2.2 Collaborative Strategies

Collaborative strategies have been advocated to improve student learning outcomes. Constructive learning theory emphasizes that learners build personal interpretations of the world based on experiences and interactions; social negotiation such as debate and discussion facilitate learning [15]. Peer instruction is one of the most collaborative methods. This approach engages students in understanding, applying, and explaining core concepts during their class [16]. Students gained more and better when taught with peer instruction than traditional methods [17].

Other approaches have also been designed to improve learning based on peer instruction. Peer-teaching has been explored to increase learning outcomes, with which one student works as an instructor to teach the fellow student(s) [18]. In the peer teaching research, students from different college years were paired in case-based learning [19].

However, little attention has been paid to improving learning outcomes with collaboration between instructors [20]. Therefore, in this study, we present a novel collaboration between two instructors. This collaboration will not increase too much work for instructors and students. Two instructors were paired in teaching to determine the impact on learning outcomes. The only requirement is that the two instructors both had experience teaching the same course.

3 Pair Teaching Method

Students' learning preferences have been categorized into different types. Learning styles are used to describe the varieties such as visual, auditory, reading/writing, and kinesthetic preferences. However, most students have more learning styles [21]. Although many researchers show learning styles are helpful for learning, there is opposition to making changes to satisfy learning styles [22].

Although there are different opinions about learning styles, and some are against others, the concept of learning styles inspires us to consider adopting the teaching preferences of instructors to improve learning outcomes. The teaching style is the pattern of attitudes and behaviors that a teacher adopts and displays during the teaching process [23]. Teaching styles may impact the learning outcomes of students [24].

Like learning styles, an instructor may apply and display more than one style in a class or different classes. In this research, we aim to utilize different teaching styles to enhance learning outcomes instead of changing the existing teaching preferences of an instructor. Most instructors agree that knowledge of learning theories is essential for instructional design. But the real benefits of that knowledge have not been realized in higher education [15]. Applying a learning theory in colleges may encounter opposition from both instructors and students. The additional work may call for resistance from teachers and students who oppose new approaches because of fear [25].

The method employed in this study is that two instructors teach a course, which is also shortly referred to as pair teaching. Instructors invited to participate in this experiment were those who had led the same lesson in the college. One reason is that each of them could take the role of the other during the procedure. Another reason is that minimal additional work is required for both teachers because they had the teaching experience in the study.

In this experiment, one teacher would take the role of another for one week in the classroom for the specified course. Both instructors would be in charge of other classes independently, and it would be difficult for them to work together all the time. Therefore, two instructors were not required to prepare the course together except that they had the same syllabus and textbook.

4 Experiment Design

4.1 Participants

Ninety-six sophomores participated in this study. Forty-two of them majored in logistics engineering, and fifty-four majored in mechanical and electronic engineering. Students were divided into two groups according to their majors. One group was chosen randomly as the control group and the other group as the treatment group. In our experiment, the control group consisted of students from the mechanical and electronic engineering major, and students who majored in logistics engineering composed the treatment group.

Two instructors joined in this study. Instructor A taught the C programming course for six years with a master's degree in mechanical engineering. Instructor A would lead the control group and design tests for both groups. Instructor B would teach the C programming course for the first time but had been teaching Java programming language course for twelve years with Ph. Degree in computer science. Instructor B would instruct the treatment group.

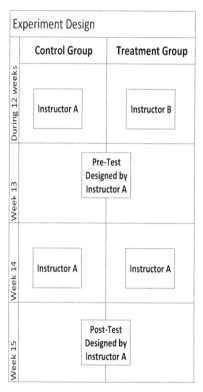

Fig. 2. Experimental design.

4.2 Course and Experiment Design

The C programming course was selected because students from both groups enrolled in it. The C course was the first computer programming course for both majors. And the C course was also the first compulsory disciplinary related course for two majors because all courses during their first year were general courses. Therefore the C programming course is suitable for the experiment.

Two groups had the same textbook, syllabus, and class schedule. Both groups had twice 2-h lectures each week for 12 weeks. In week 13, students learning outcomes were assessed with the pre-test designed by Instructor A. In week 14, besides teaching the control group, instructor A instead of instructor B led the treatment group. In week 15, students who failed in the pre-test from both groups took the post-test. The post-test was also designed by instructor A. The experiment design is shown in Fig. 2.

4.3 Data Analysis Methods

The grades of the pre-test from the two groups would be analyzed with nonparametric two independent samples to determine if there is a statistical difference between the means for the control group and the treatment group.

As for the post-test grades, nonparametric independent samples test methods would be employed to verify whether there is a significant difference between the means of the two groups because of the small sample size.

5 Results

Forty-two sophomores majoring in logistics engineering formed the treatment group. The control group consists of fifty-four students majoring in mechanical and electronic engineering. The instructor (Instructor A) in charge of the control group had six years of experience in teaching the C course. Another teacher (Instructor B) lecturing the treatment group taught the C course for the first time. The pre-test and post-test were both designed by Instructor A alone.

Both instructors taught their group the course independently for fourteen weeks. Then students from both groups took the pre-test. The frequency distribution of grades for the control group is shown in Fig. 3. And the frequency distribution of scores from the treatment group is shown in Fig. 4.

The descriptive statistics for the control group are: $n = 54$, min $= 27$, max $= 90$, mean $= 65.56$, sd $= 18.15$, and $p = 0.004$ with Shapiro-Wilk for normalization test. And the descriptive statistics for the treatment group are: $n = 42$, min $= 10$, max $= 68$, mean $= 39.95$, sd $= 15.63$, and $p = 0.152$ with Shapiro-Wilk for normal distribution test. Therefore, the data from the control group do not follow normal distribution.

There is a significant statistical difference between the means of the two groups for the pre-test with the nonparametric independent-samples Mann-Whitney Test (with SPSS 26, $N = 96$, Test Statistic $= 1931.500$, and 2-tailed $p = 0.00 < 0.05$). Therefore, students achieved a higher mean of grades for the pre-test in the control group than in the treatment group.

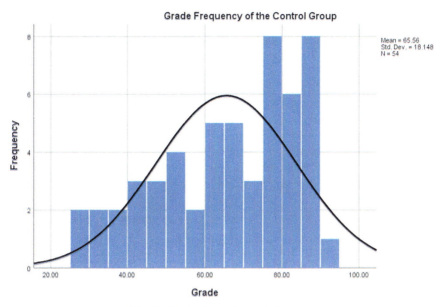

Fig. 3. Grades from the control group.

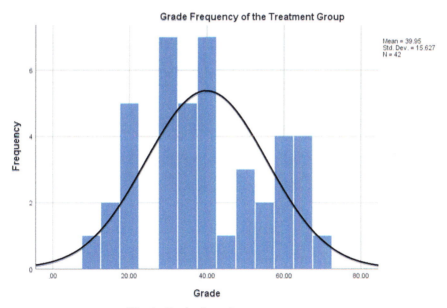

Fig. 4. Grades from the treatment group.

Then Instructor A lectured both groups separately for one week. In week 15, students who had grades in the pre-test less than sixty from both groups took the post-test. Seven students in the control group and twenty-six students in the treatment group took the post-test. The frequency distribution of post-test grades for the control group is shown in Fig. 5, and the frequency distribution for the treatment group is shown in Fig. 6.

The descriptive statistics of post-test grades for the Control Group are: n = 7, min = 38, max = 67, mean = 52.1, sd = 11.14 and for the Treatment Group are: n = 16, min = 62.9, max = 97.2, mean = 81.0, sd = 10.20.

There is a significant statistical difference between the average grades of the two groups for the post-test (N = 33, 2-sided test p = 0.00 < 0.05, the nonparametric independent samples test with SPSS 26).

Therefore, students got a higher mean of grades for the post-test in the treatment group than in the control group. Students gained more in the treatment groups during the pre-test and the post-test than in the control group.

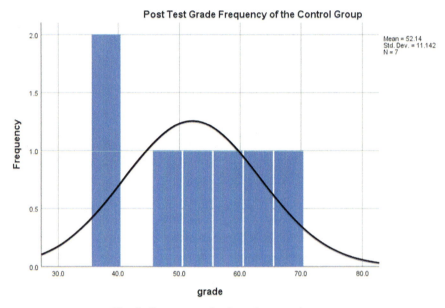

Fig. 5. Post-test grades from the control group.

6 Discussion

Students from the control group achieved a higher average score (mean = 65.56) in the pre-test than students did (mean = 39.95) from the treatment group. When we interviewed students in the treatment group, most of them said they did not practice those questions on the test before. One contribution is possible from the similarity between the context students learned, and the context students took the test. Instructor A taught the control group and also designed the pre-test. Therefore, students could be familiar with the teaching styles of Instructor A.

Interestingly, students in the treatment group obtained higher scores (mean = 81.10) in the post-test than students did (mean = 52.14) in the control group. We might attribute

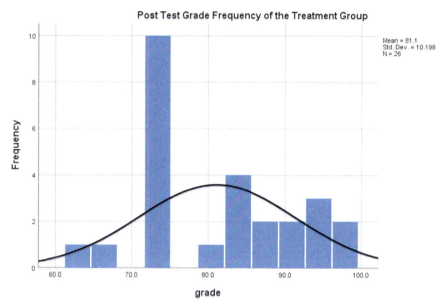

Fig. 6. Post-test grades from the treatment group.

this radical change to the pair teaching strategy. After Instructor A had taught the treatment group for a week, students in the treatment groups got to know the teaching styles of Instructor A.

Students in the treatment group achieved better outcomes in the post-test than in the control group. The pair teaching strategy has improved the learning outcomes. The cognitive learning theory could explain this phenomenon, the ability in knowledge transfer [26]. Thus, the pair teaching method conforms to cognitive learning theories. The strategy presented in this research is different from the collaborative but more effort needed process, which formed a learning improvement team to help faculty [27]. This pair teaching method is also different from another collaborative approach between instructors of related courses [20]. That approach focuses on planning and designing curriculum content.

However, this research has some limitations. The sample size in the control group for the post-test is small (N = 7). In the future, more experiments should be designed to verify the learning outcomes improvement of the pair teaching method. Another limit is that instructor A participated in the teaching process in the treatment group only in the last week. Is it possible that students in the treatment group could achieve more in the pre-test if the instructor had adopted the pair teaching method from the beginning? Therefore, more work should be done to answer the question.

7 Conclusion

Two instructors collaborated to teach one course in higher education with the requirement that both have taught or can teach the same course. The pair leading strategy demands no more additional work than a traditional approach, which encourages instructors to participate in the teaching activities. The results from the experiment show that students achieved better learning outcomes with pair teaching than with the traditional method. The results indicate that the pair teaching strategy can improve students learning outcomes in higher education.

Acknowledgment. Yongbin Zhang thanks the Beijing Municipal Education Commission and the Ministry of Education of the People's Republic of China (202102122003) for supporting this research.

References

1. Caspersen, J., Smeby, J.C., Olaf Aamodt, P.: Measuring learning outcomes. Eur. J. Educ. **52**(1), 20–30 (2017)
2. Harden, R.M.: Developments in outcome-based education. Med. Teach. **24**(2), 117–120 (2002)
3. Wolf, R., Zahner, D., Benjamin, R.: Methodological challenges in international comparative post-secondary assessment programs: lessons learned and the road ahead. Stud. High. Educ. **40**(3), 471–481 (2015)
4. O'Keeffe, L., O'Halloran, K.L., Wignell, P., Tan, S.: A linguistic analysis of the sample numeracy skills test items for pre-service teachers issued by the Australian council for educational research (ACER). Aust. Educ. Res. **44**(3), 233–253 (2017)
5. McClellan, J.: Beyond student learning outcomes: developing comprehensive, strategic assessment plans for advising programmes. J. Higher Educ. Policy Manage. **33**(6), 641–652 (2011)
6. Keshavarz, M.: Measuring course learning outcomes. J. Learn. Des. **4**(4), 1–9 (2011)
7. Wieman, C.E., Adams, W.K., Perkins, K.K.: PHYSICS PhET: simulations that enhance learning. Science **322**(5902), 682–683 (2008)
8. Monteiro, S., Sibbald, M.: Aha! taking on the myth that simulation-derived surprise enhances learning. Med. Educ. **54**(6), 510–516 (2020)
9. Deslauriers, L., Schelew, E., Wieman, C.: Improved learning in a large-enrollment physics class. Science **332**(6031), 862–864 (2011)
10. Duraes, D., Toala, R., Goncalves, F., Novais, P.: Intelligent tutoring system to improve learning outcomes. AI Commun. **32**(3), 161–174 (2019)
11. Schreurs, J., Dumbraveanu, R.: A shift from teacher centered to learner centered approach (in en). Int. J. Eng. Pedagogy (iJEP) **4**(3), 36–41 (2014)
12. Taylor, R.M.: Defining, constructing and assessing learning outcomes. Revue Sci. Et Tech.-Off. Int. Des Epizooties **28**(2), 779–788 (2009)
13. Driscoll, A., Wood, S.: Developing Outcomes-Based Assessment for Learner-Centered Education: A Faculty Introduction, 1st edn, p. 275. Stylus, Sterling, Va (2007)
14. Harden, R.M.: Learning outcomes and instructional objectives: is there a difference? Med. Teach. **24**(2), 151–155 (2002)
15. Ertmer, P.A., Newby, T.J.: Behaviorism, cognitivism, constructivism: comparing critical features from an instructional design perspective. Perform. Improv. Q. **26**(2), 43–71 (2013)
16. Crouch, H., Mazur, E.: Peer instruction: ten years of experience and results. Am. J. Phys. **69**(9), 970–977
17. Lasry, N., Mazur, E., Watkins, J.: Peer instruction: from Harvard to the two-year college. Am. J. Phys. **76**(11), 1066–1069 (2008)

18. Nshimiyimana, A., Cartledge, P.T.: Peer-teaching at the University of Rwanda—a qualitative study based on self-determination theory. BMC Med. Educ. **20**(1), 230 (2020)
19. Oh, S.-L., Jones, D.: Sophomore-senior case studies: a case-based learning exercise with peer teaching. J. Dent. Educ. (2020)
20. Duke, J.M., Awokuse, T.O.: Assessing the effect of bilateral collaborations on learning outcomes. Rev. Agric. Econ. **31**(2), 344–358 (2009)
21. Samarakoon, L., Fernando. T., Rodrigo, C., Rajapakse, S.: Learning styles and approaches to learning among medical undergraduates and postgraduates. Bmc Med. Educ. **13**(42) (2013)
22. Kirschner, P.A.: Stop propagating the learning styles myth. Comput. Educ. **106**, 166–171 (2017)
23. Hurriyetoglu, N., Kilicoglu, E.: Teaching styles scale: validity and reliability study. Educ. Policy Anal. Strateg. Res. **15**(2), 222–237 (2020)
24. Inayat, A., Ali, A.Z.: Influence of teaching style on students' engagement, curiosity and exploration in the classroom. J. Educ. Educ. Dev. **7**(1), 87–102 (2020)
25. Weimer, M.: Learner-Centered Teaching: Five Key Changes to Practice, 2nd ed. Jossey-Bass (2013)
26. Palincsar, S.: Social constructivist perspectives on teaching and learning. Ann. Rev. Psychol. **49**(1), 345 (1998)
27. McMillan, L., Johnson, T., Parker, F.M., Hunt, C.W., Boyd, D.E.: Improving student learning outcomes through a collaborative higher education partnership. Int. J. Teach. Learn. Higher Educ. **32**(1), 117–124 (2020)

Foreign Language Reading Anxiety and Its Correlation with Reading Test Scores

Van T. T. Dang and Trung Nguyen(✉)

FPT University, Hanoi, Vietnam
{vandtt3,trungnb6}@fe.edu.vn

Abstract. Anxiety is a common experience among all of us in every daily situation. It is considered one of the affective factors in foreign language learning. Past research has pointed out that foreign language anxiety negatively correlates with academic performance in all four traditional language skills (speaking, listening, writing, and reading). Of all the skills, reading has been once deemed less susceptible to anxiety; however, it has been increasingly gaining more attention from researchers since findings of the pioneers (Saito and the associates) as a phenomenon related to foreign language anxiety. Notwithstanding increasing studies on foreign language reading anxiety, not much has been conducted in Vietnamese contexts, especially the research site. Thus, a survey design was conducted on 207 advanced level students at an educational Vietnamese institution with the aim to explore participants' reading anxiety level, possible differences between FLRA and demographic variables, and correlation between FLRA and reading test scores. Results revealed that 96.6% of respondents were at a moderate level of reading anxiety. There were no statistically significant differences between FLRA and demographic variables such as gender, birthplace, and English text reading time per week. Similarly, no statistically significant correlation between FLRA and reading test scores could be found. Pedagogical implications were also discussed.

Keywords: Foreign Language Reading Anxiety (FLRA) · Reading test scores · FLRA level · Reading time

1 Introduction

Anxiety, defined as "the subjective feeling of tension, apprehension, nervousness, and worry associated with an arousal of the autonomic nervous system" by Spielberger (as cited in [1], p. 113), is believed to hamper the process of learning a foreign language [1]. Under psychological views, anxiety encompasses three main types: Trait anxiety, state anxiety, and situation-specific anxiety. The first one, a general personality trait, is stable over time and can affect cognition and memory. State anxiety refers to apprehension that is felt at a specific point in time. It is temporary anxiety experienced on a moment-by-moment basis and is linked to specific occasions or circumstances [2] while anxiety triggered by a specific scenario or event, such as public speaking or class participation, is referred to as situation-specific anxiety [3]. It is also held that language learning is

© The Author(s), under exclusive license to Springer Nature Singapore Pte Ltd. 2023
E. C. K. Cheng et al. (Eds.): AIET 2022, LNDECT 154, pp. 168–181, 2023.
https://doi.org/10.1007/978-981-19-8040-4_13

related to situation-specific anxiety other than other types since language learners suffer from anxiety in various situations in a language class such as listening, speaking, test, public speaking, writing, reading and so on.

Much research has examined and evaluated foreign language listening / speaking / writing anxiety while research on reading anxiety has gained more attention recently. Evidenced established that foreign language learners experience reading anxiety, which has been proved via findings on Foreign Language Reading Anxiety (FLRA) levels (namely high-anxiety, mid-anxiety, and low-anxiety) [4, 5]. In addition, major factors contributing to reading anxiety are found to be text features (including unknown vocabulary, unfamiliar topics, unknown grammar, and unfamiliar culture) and personal features (e.g., fear of making mistakes, worry about reading effect, lack of concentration and so on) [6, 7, 8, 9]. The inevitable besides causes of FLRA is its effects on reading strategies, reading process, reading comprehension, reading performance and so forth [10, 11, 12, 13, 14]. Correlations between FLRA and demographic variables (e.g., gender, language proficiency, educational levels) were also found [15, 16, 5, 17, 18].

Notwithstanding growing literature on FLRA regionally and worldwide, very limited published studies are found in Vietnamese contexts; especially no research has been conducted in the research site. Additionally, the present research expects to extend the existing literature on the relationship between FLRA and demographic information (such as birthplace and time spent on reading English texts) and reading test scores.

1.1 Purpose and Significance of the Research

The aims of the current research were threefold. The primary one was to explore students' level of FLRA. The study also sought to examine the relationship between demographic variables and FLRA and the correlation between FLRA and reading test scores. When these aims were served, results expectantly help English instructors have clear insights into how anxious students are, hence devising effective teaching methods to alleviate students' reading anxiety.

1.2 Research Questions

With the aforementioned aims in mind, the current research attempts to answer the following research questions:

1. What is Advanced Students' Level of Foreign Language Reading Anxiety?
2. How is foreign language reading anxiety related to demographic variables, specifically gender, birthplace, and time spent on reading English texts weekly?
3. Is There a Relationship Between FLRA and Reading Test Scores?

2 Literature Review

2.1 Foreign Language Reading Anxiety

Saito et al. pioneered to conduct research on FL reading anxiety and concluded that reading apprehension is "a phenomenon related to, but distinct from, general FL anxiety" [19]

(p. 211). Their conclusion has opened multiple subsequent research on various aspects associated with reading anxiety. To make it clearer, Zhao et al. defined this anxiousness as the sentiments of dread and worry learners have when reading a target language [20]. Under psychological views, Zbornik (as cited in [21], p. 343) referred it to "a specific, situational phobia toward the act of reading that has physical and cognitive reactions". Most studies have considered it as cognitive reactions which have a connection with thinking or mental processes other than physical ones including sweating, feeling shaky or faint, pulse rate and so forth. In the current research, FL reading anxiety is considered to influence cognition; in other words, it is worry and uneasiness during the reading process.

2.2 Foreign Language Reading Anxiety Level

The first dimension of reading anxiety revealed by past research is the level of anxiety. Sellers, followed by Gonen, proposed a formula calculating the level based on Mean and Standard Deviation [4, 5]. Accordingly, they categorized three groups of anxiety level as follows:

High anxiety: Mean + Standard deviation = The score higher than this.

Low anxiety: Mean – Standard Deviation = The score lower than this.

Medium Anxiety: The score between Mean – Std. Deviation and Mean + Std. Deviation.

Building on this form, following researchers have demonstrated flexible level of anxiousness in their research participants. Respondents in Sabti et al., Bensalem, and Abubakar and Hairuddin confirmed their high anxiety with the percentage of 52.4%, 65.63%, and 87.3% respectively [22, 23, 24]. Medium anxiety were reported in Razak et al., Muhlis, Isler and Yildirim, Wijayati et al., and Al-Sohbani [25, 26, 6, 27, 28] while Faruq [29] indicated low anxiety level among respondents. Beyond the formula, Lu and Liu [30] disclosed no anxiety when reading among more than 50% of research participants. Notably, Gonen [5] uncovered that the more English proficient students are, the lower anxiety they experience when investigating three groups of language proficiency students (e.g., Elementary, Intermediate, and Advanced English proficiency).

2.3 FLRA and Correlates

Past studies have illustrated FL reading anxiety in the relationships or correlation with distinctive variables. For the most part, **demographic variables** have been recorded to be in relation with reading apprehension. One of the prevalent demographic information revealing conflicting associations is gender [22, 23, 31, 15, 32, 30]. Anxiety was reported to affect males rather than females, which means males are more anxious when reading than female counterparts [30]. In some other cases, female participants experienced higher anxiety than the opposites [23, 31, 15, 22]. Likewise, Tien [17] held that females appeared to be more anxious in some aspects of reading than males. Specifically, they found new English symbols more difficult to them when reading or they admitted reading to be the hardest part to learn. In contrast, Capan and Karaca [18] and Capan and Pektas [11] showed no interrelationship between FL reading anxiety and gender. In alignment with these two studies, Tsai and Lee [8], Chow et al. [10], and Matsuda and Gobel [32] suggested no indication of gender differences in reading trait anxiety.

Another demographic variable correlated with reading apprehension is language proficiency. Capan and Karaca [18] found that the least anxious group is the most proficient students. In a like manner, Gonen [5] maintained that there is a positive relationship between reading anxiety and language competence, meaning that anxiety decreases if students are more proficient in FL learning. For instance, elementary learners in Gonen's study [5] suffered from more anxiety than other proficiency level students. Having the same notion, Lien [16] detected that self-perceived reading proficiency is significantly correlated with anxiety. Those with a higher level of English felt more confident and satisfied with reading process than those with lower levels. However, these findings do not always hold true, which means it is not always that the more English proficient learners are, the less anxious they feel. Evidence established that there were no significant differences in reading anxiety among less competent learners, average readers, and more proficient [17].

In addition to these most prevailing demographic variables, others have also been examined. Travel experience in countries whose language learners are learning and knowledge of the third language are found to have a small effect size, i.e., a small relationship with reading apprehension [23]. In terms of time spent on reading English text weekly, Tien [17] uncovered no significant correlation with FL reading anxiety. When it comes to grade levels, Capan and Karaca [18] indicated a moderately significant difference in reading anxiousness between second-year and third-year students, and there was a nonlinear decrease in anxiety level in relation to the number of years spent in school.

As an **independent variable**, FL reading anxiety has been proved to have an impact on reading comprehension and the use of reading strategies. For the first part, most research found negative or no significant correlation between FLRA and reading comprehension. For instance, Chow et al. [10] maintained that reading anxiety, as measured by both reading trait and state anxiety, negatively linked with ESL reading comprehension with a moderate correlation. Having the same conclusion, Gaith [13] reported FLRA and reading comprehension have a direct negative statistically significant association with $r = -.28, p < .01$. Data from Mardianti et al. [27] also revealed a negative moderate-strong connection between ESP students' reading anxiety and comprehension, meaning that the more anxious students are, the lower their understanding score. Further, Nazarinasab et al. [33] sought to evaluate how foreign language reading anxiety affected ESP text comprehension of university students and unveiled a negative significant relationship between the two variables. This shows that when FLRA is reduced, reading comprehension improves. By contrast, Brantmeier [14] examined the relationship between anxiety levels and post-written tasks of reading comprehension and found no positive correlation between them. Finding the same results, Hassaskhah and Joghataeian [34] also demonstrated no statistically significant correlation between reading comprehension test scores of advanced learners and FLRA. With respect to the interrelationship between reading apprehension and reading strategies, Lien [15] discovered that more anxious learners utilized fewer reading strategies. Specifically, males are less anxious than females and tend to use more reading strategies than their counterparts who experienced more anxiety.

Conversely, Ghonsooly and Loghmani [35] reported no significant correlation between EFL learners' FLRA and their use of reading strategies.

As a **dependent variable**, critical thinking, text difficulty, and reading strategies may be predictors of FL reading anxiety. Capan and Pektas [11] conducted an experiment to see if there are any changes in FLRA between the experiment and control group. Results showed that after receiving training on various reading strategies such as skimming, scanning, maintaining reading logs, and reflective thinking, FLRA levels of individuals in the experiment group increased while mean scores of pre- and post-tests by the control group were not statistically significant. Another experiment with two groups ($i-1$ group who were assigned graded readers stories below their level of proficiency and $i + 1$ group who were assigned stories beyond their level of proficiency) by Bahmani and Farvardin [12] suggested that the "$i + 1$" group's FLRA was significantly improved at the end of the extensive reading training while the "$i-1$" group's FLRA was significantly decreased after the intervention. Considering critical thinking skills, Aghajani and Gholamrezapour [36] proposed when critical thinking skills improve, reading anxiousness decreases. Moreover, Genc [7] found out that second language ambiguity tolerance, success in reading in a foreign language, and gender could be significant predictors of reading anxiety in a target language, implying that ambiguity tolerance employment could alleviate the apprehension.

2.4 Contributions and Gaps

Regarding contributions, it is undeniable that findings from these studies have made a great contribution to the existing literature, which has changed past researchers' perspectives that reading is less susceptible to anxiety than other skills. Indeed, researchers have calculated reading anxiety levels, which indicated its existence and functioning as an independent issue. Further, reading anxiousness is also believed to have a correlation with multiple variables from learners such as gender, language proficiency, the use of reading strategies, reading comprehension, reading performance, and so on. Notwithstanding these contributions, its correlations with birthplace and reading test scores seem unavailable, and relationships with time spent on reading is still very limited. More importantly, research on reading anxiety in the Vietnamese contexts is possibly rare since few published research has been found, except one on speaking anxiety by Le [37] and ones on foreign language anxiety by Tran [38, 39, 40]. In the researched site, no research has been conducted on this issue, to my knowledge. These gaps motivate the researcher and lay the foundation for the current study.

3 Methodology

3.1 The Survey Design

A cross-sectional survey was employed with the primary aims to disclose students' level of Foreign Language Reading Anxiety (FLRA), relationships between demographic variables and FLRA, and possible correlation between FLRA and reading test scores. The selected design is beneficial and convenient for the current research since utilizing a hyperlink of Google Form other than paper questionnaires is economical and rapid to collect a dozen of responses free of charge. The Google Form also reduced constraints on data entry and data missing processing which could be avoided when all questions are set to be compulsorily ticked.

3.2 Sample

Research participants encompassed 207 out of approximately 825 Advanced level students in the Preparational English course. Their proficiency of English (Advanced level) was determined by a Placement test and they were taking a reading skill-specific course titled Advanced Reading Success at the time of the current research. More males took part in the research than females with the percentage of 69.1% and 30.9% respectively. More than half (51.2%) originate from urban areas while 48.8% come from rural areas. Nearly a third (32.9%) are majoring in Business Administration whereas only 3.9% are language-majored students. Over sixty-three percent of respondents are in Information Technology majors.

3.3 Instrumentation

The research instrument is a questionnaire survey including two main parts. Part 1 is Foreign Language Reading Anxiety Scale developed by Saito et al. [19]. It is five-point Likert scale 20 items to "elicit students' self-reported anxiety over various dimensions of reading, their perceptions of reading difficulties, and their perceptions of the relative difficulty of reading as compared to the difficulty of other language skills" (p. 204). Distinct from Saito et al.'s study which conducted on students of French, Japanese, and Russian language, the current one centered on English foreign language. Also, in contrary to the original scale whose ranging scores is from 20 to 100, the adapted one's scores range from 1 (*strongly disagree*) to 5 (*strongly agree*) implying that the higher the score, the more anxious students are. Part 2 is demographic information requiring informants to provide data on their gender, majors, birthplace, time spent on reading English texts every week, and their latest reading test scores. The one related to test scores is a short answer where participants write reading test scores on their own.

3.4 Procedures

Thanks to teacher supports, an email with Google Form hyperlink was forwarded to 12 intact classes (approximately 300 students of English Advanced level). Two hundred and twenty responses were returned with a response rate of over 73.3%. Data were then downloaded and coded. Coding allowed the researcher to eliminate bias responses such as identical options for all questions, unknown reading test scores, and missing data. Subsequently, data were imported into SPSS version 26 for analyses.

4 Results

4.1 Research question 1: What is advanced students' level of Foreign Language Reading Anxiety?

Scale scores range from 1 to 5, with point 3.0 assigned neutral or neither agree nor disagree, and mean scores of 20 FLRA items were mainly over 3, indicating that most respondents agree with items of FLRAS. This also pinpoints that they were moderately anxious when reading English texts. Subsequently, average mean and standard deviation of 20 items was calculated, and FLRA mean score of each participant was also computed prior to determining anxiety level. Results suggested Mean = 3.17 and Std. Deviation = 0.98. Based on the formula by Seller [4] and Gonen [5], high-anxiety is gauged by sum of Mean and Std. Deviation (3.17 + 0.98 = 4.16) while low-anxiety is the result of Mean subtracting from Std. Deviation (3.17 – 0.98 = 2.19). Mid-anxiety lies in between 2.19 and 4.16. As indicated in Table 1, 200 out of 207 respondents (equivalent to 96.6%) belong to mid-anxiety while only three (1.4%) and four (1.9%) are low- and high-anxiety respectively.

Table 1. FLRA level groups.

FLRA level groups	Frequency	Valid percent
High-anxiety	4	1.9
Mid-anxiety	200	96.6
Low-anxiety	3	1.4
Total	207	100

4.2 Research Question 2: Is There a Difference Between Foreign Language Reading Anxiety and Demographic Variables, Specifically Gender, Birthplace, and Time Spent on Reading English Texts Weekly?

In the realm of the difference in FLRA between males and females, whether scores of FLRA normally distributed for each group was evaluated prior to statistical techniques to compare groups. The result of the Kolmogorov-Smirnov statistic suggested violation of the assumption of normality with the Sig. Value being 0.001. Thus, the researcher utilized the Mann-Whitney U Test for differences between two independent groups (males and females) rather than t-test. As a result, the Mann-Whitney U Test revealed no significant difference in the FLRA scores of males ($Md = 3.15, n = 143$) and females ($Md = 3.225, n = 64$), $U = 4114.000, z = -1.161, p = .245$.

With respect to the difference in FLRA between those born in rural and urban areas, Mann-Whitney U Test was conducted once more. The results indicate non-significant differences between rural ($Md = 3.2, n = 101$) and urban groups ($Md = 3.15, n = 106$), $U = 4921.000, z = -1.004, p = .315$.

When it comes to differences in FLRA scores between groups of reading time, descriptive statistics were first produced. As presented in Table 2, more than half of respondents (51.2%) were reading English texts below one to three hours per week, followed by the group *over three to six hours* (over a quarter). Only 10 participants informed no reading time while 16 read over nine hours per week.

Table 2. Frequencies of reading time.

Reading time	Frequency	Valid percent
Group 1 (0 h)	10	4.8
Group 2 (Below 1 – 3 h.)	106	51.2
Group 3 (Over 3 – 6 h.)	52	25.1
Group 4 (Over 6 – 9 h.)	23	11.1
Group 5 (Over 9 h.)	16	7.7

A one-way ANOVA was performed to compare the effect of reading time on FLRA scores. Participants were divided into five groups according to their time spent on reading English texts (Group 1: 0 h; Group 2: Below 1–3 h; Group 3: Over 3–6 h; Group 4: over 6–9 h; Group 5: Over 9 h). Results revealed that there was no statistically significant difference in mean FLRA scores for the five reading time groups: $F(4, 202) = .494$, $p = .74$ (see Table 3).

Table 3. One-way ANOVA results.

FLRA	Sum of squares	df	Mean square	F	Sig.
Between groups	.312	4	.078	.494	.740
Within groups	31.925	202	.158		
Total	32.237	206			

4.3 Research question 3: Is there a relationship between FLRA and reading test scores?

The question on the recent reading test score is in the form of a short answer where students write scores on their own. To explore the relationship between this variable with FLRA scores, the researcher first converted reading test scores into five groups, based on Vietnam's grading system (Group 5: 8.5–10; Group 4: 7–8.4; Group 3: 5.5–6.9; Group 2: 4.0–5.4; Group 1: < 4.0). Descriptive statistics was also performed to have insights into the score ranges. Table 4 shows that score ranges of group 1, 3, and 4 are not too far different with 17.9%, 16.4% and 14.5% respectively. The highest scores lie in group 5

Table 4. Reading test score descriptive statistics.

Score ranges	Frequency	Valid percent
Group 1: <4	37	17.9
Group 2: 4 – 5.4	24	11.6
Group 3: 5.5 – 6.9	34	16.4
Group 4: 7 – 8.4	30	14.5
Group 5: 8.5 –10	82	39.6
Total	207	100

with the percentage of nearly 40%; it means 82 out of 207 respondents confirmed their reading test scores were between 8.5 and 10.

As mentioned in research question 2, the distribution of FLRA scores violated the assumption of normality; therefore, Spearman's Rank Order Correlation was performed. Results indicated no statistically significant correlation between the two variables, $r = -.079, n = 207, p > .255$ (see Table 5).

Table 5. Correlation between FLRA scores and reading test scores.

		FLRA scores
Reading test scores	Spearman's rho	–.079
	Sig. (2-tailed)	.255

5 Discussion

The first finding of the current research is the moderate level of FL reading anxiety in advanced students with the percentage of 96.6%. This is a reconfirmation of FLRA existence and susceptibility to anxiety of reading skill. However, research participants are expected to experience low-anxiety, not mid-anxiety, since they are advanced proficiency students. According to CEFR, advanced students are C1 level, and this level requires learners to be able to understand lengthy and demanding texts and recognized indirectly stated meaning. Further, all topics of the coursebook (Reading Advanced Transition) are suited for C1 level students as suggested by editors of the book. So, why are they moderate in anxiety? The answer could be the time of training course. It is 105 h in seven weeks for one course which seems insufficient to level up. CEFR also says it should take 200 h to reach B2 from B1 and C1 from B2.

The finding is supported by those from past research including Muhlis [26], Isler and Yildirim [6], Bensalem [23], Faruq [29], Razak et al. [25], Mardianti et al. [27], and Al-Sohbani [28], who also calculated anxiety levels based on Mean and Standard Deviation and asserted mid-anxiety in research respondents. Nevertheless, the percentage of mid-anxiety student in the current study outperforms that of the others with 96.6% compared with 71.9%, 90%, 62%, and 44%. Further, the study was conducted on students of Business Administration and Information Technology in comparison with students of grade 11 in Muhlis's [26], students studying English language teaching program in Isler and Yildirim's [6], and Governmental students in Mardianti et al.'s [27].

The second finding suggests no statistically significant differences between FLRA scores and demographic variables such as gender, birthplace, and total weekly reading time. In relation to males and females, non-significant difference has also been found in studies by Capan and Karaca [18], Capan and Pektas [11], Tsai and Lee [8], Chow et al. [10], and Matsuda and Gobel [32]. This seems hold true as pinpointed by Huberty [41] (p. 1) that "anxiety is a common experience to all of us". It is arduous to say whose anxiety outdoes the others. Regarding the birthplace variable, no research has evaluated it on reading anxiety levels. The finding on the last researched demographic variable, weekly reading time, is in alignment with that of Tien [17]. It could be explained that they might have read in English for pleasure, and if that works, reading materials are possibly easier than those from coursebooks that covers subjects such as Linguistics, Earth Science, Business Ethics, Medieval Culture, and Materials Engineering. These topics are academic, and rare learners read them for pleasure. Thus, total reading time per week between groups could not make a difference in reading anxiety.

The last finding is no statistically significant correlation between FLRA scores and reading test scores. It could be that reading scores play a small part in total scores of a test. In other words, learners had to take two progress tests during the reading skill-specific course, and the final score is calculated by sum of reading score (40%), vocabulary score (20%), writing score (20%), class activity score (10%) and workbook score (10%) while final examination score is the sum of reading, vocabulary and writing. This shows that students appear not to be so anxious in reading English texts since they might believe they could pass the course thanks to other test parts (vocabulary, and writing). This finding is not supported by any previous evidence because no similar findings could be found.

Contributions of the present research lie in the fact that no similar studies have been conducted in the research site, and few published ones could be found in the Vietnamese contexts. This helps add the local research site to the existing literature. Furthermore, results of birthplace, reading time, and reading test scores in relation to FL reading anxiety could extend the literature. No or not much research has evaluated these relationships.

6 Conclusions

The aim of the research is to explore (1) learners' FL reading anxiety level, (2) possible differences between FLRA and demographic variables, and (3) correlation between reading anxiety and reading test scores. Results revealed that research participants experienced a moderate level of anxiety with the percentage of 96.6%. However, there were no statistically significant differences between FL reading anxiety and demographic variables. A similar finding was found in relationship between FLRA and reading test scores. What have been found not only adds to but also extends the existing literature.

Results of the research should be also analyzed in light of limitations. In terms of participants, the researcher expected to obtain responses from all advanced learners; however, this was hindered due to school closure during the Covid-19. Consequently, only 220 out of approximately 825 responses were returned. Only a fourth doing the survey seems inadequate to make generalizations as suggested by Creswell and Guetterman [42] that the number of survey participants should be approximately 350. The hinderance also affected the selection of research approach for the study. Initially, the researcher intended to conduct a mixed methods study, but it was changed because of Covid-19 pandemic. Thus, true reasons behind their anxiety level, time they spent on reading English texts weekly, and reading test scores cannot be revealed.

Based on the results and limitations, some further research is suggested. A mixed methods research study is a better selection as it would provide a full understanding of the research problem. Specifically, what made them have high reading test scores while they are moderate in reading anxiety? Is it because anxiety is functioning positively, a facilitating anxiety type? Why they are in mid-anxiety, not low-anxiety as expected for advanced level students is another concern. Or what reading materials they read during the week could help researchers evaluate better the relationship between FLRA and reading time. Moreover, the number of research participants should be more than 207, at least 350 as suggested by Creswell and Guetterman [42]. Lastly, what reading strategies learners are utilizing should be researched to see whether there is any correlation with or their effects on FLRA since they are taking reading skill-specific course.

Given the study findings, the following implications are made for faculties to help their learners reduce reading anxiety. First and foremost, teaching reading strategies could lower the anxiety and consequently increase students' reading comprehension. It is fortunate that the coursebook *Advanced Reading University Success*, an academic course designed for English language learners preparing for mainstream academic environments, covers not only fundamental reading skills but also critical thinking skills that brings students a comprehensive insight into what skills are needed for rapid and accurate reading comprehension. Teacher roles are not only teaching these skills but also encouraging and testing students' skills application. Justification is that from theory to practice is a long way. Students understand skills but how they apply them is another question. From the researcher's observation, learners seem to grasp knowledge rapidly but slow in practicing it. Moreover, academic content of topics such as Earth Science, Linguistics, Business Ethics, Materials Engineering, and Medieval Culture is totally new and sometimes boring to students; thus, redesigning lessons is crucial. For example, traditional lectures could be replaced with constructivism types that would aid students themselves in exploring the importance of reading skills and academic content. Another way is students could become guest speakers to present about topics on their own. Teacher could provide learners with more materials related to the topics. The more exposure to academic topics they receive, the more familiar they feel and less anxious they are during the reading process.

References

1. Horwitz, E.K., Horwitz, M.B., Cope, J.: Foreign language classroom anxiety. Mod. Lang. J. **70**(2), 125–132 (1986). https://doi.org/10.1111/j.1540-4781.1986.tb05256.x
2. Dörnyei, Z.: Attitudes, orientations, and motivations in language learning: advances in theory, research, and applications. Lang. Learn. **53**(S1), 3–32 (2003). https://doi.org/10.1111/1467-9922.53222
3. Horwitz, E.: Language anxiety and achievement. Annu. Rev. Appl. Linguist. **21**, 112–126 (Jan. 2001). https://doi.org/10.1017/S0267190501000071
4. Sellers, V.D.: Anxiety and reading comprehension in Spanish as a foreign language. Foreign Lang. Ann. **33**(5), 512–520 (2000). https://doi.org/10.1111/j.1944-9720.2000.tb01995.x
5. Gonen, S.I.K.: L2 reading anxiety: Exploring the phenomenon, p. 10
6. Isler, C., Yildirim, O.: Sources of turkish efl learners' foreign language reading anxiety. (2017). doi: https://doi.org/10.30762/JEELS.V4I1.328
7. Genç, G.: Can ambiguity tolerance, success in reading, and gender predict the foreign language reading anxiety?. J. Lang. Linguist. Stud., **12**(2), Art. no. 2, (Oct. 2016)
8. Tsai, Y.-R., Lee, C.-Y.: An exploration into factors associated with reading anxiety among taiwanese efl learners. TEFLIN J.—Publ. Teach. Learn. Engl., **29**(1), p. 129 (Jul. 2018), doi: https://doi.org/10.15639/teflinjournal.v29i1/129-148
9. Sheikh Ahmad, I., Al-Shboul, M.M., Sahari Nordin, M., Abdul Rahman, Z., Burhan, M., Basha Madarsha, K.: The potential sources of foreign language reading anxiety in a Jordanian EFL context: A theoretical framework. Engl. Lang. Teach., **6**(11), p. p89 (Oct. 2013), doi: https://doi.org/10.5539/elt.v6n11p89
10. Chow, B., Mo, J., Dong, Y.: Roles of reading anxiety and working memory in reading comprehension in english as a second language. Learn. Individ. Differ. **92**, 102092 (Oct. 2021). https://doi.org/10.1016/j.lindif.2021.102092
11. Ahmet Capan, S., Pektas, R.: An empirical analysis of the relationship between foreign language reading anxiety and reading strategy training, Engl. Lang. Teach., **6**(12), p. p181 (Nov. 2013), doi: https://doi.org/10.5539/elt.v6n12p181
12. Bahmani R., Farvardin, M.T.: Effects of different text difficulty levels on EFL learners' foreign language reading anxiety and reading comprehension, p. 18
13. Ghaith, G.M.: Foreign language reading anxiety and metacognitive strategies in undergraduates' reading comprehension, p. 19
14. C. Brantmeier, "ANXIETY ABOUT L2 READING OR L2 READING TASKS? A STUDY WITH ADVANCED LANGUAGE LEARNERS," p. 19
15. Lien, H.-Y.: EFL Learners' reading strategy use in relation to reading anxiety. Lang. Educ. Asia **2**(2), 199–212 (Dec. 2011). https://doi.org/10.5746/LEiA/11/V2/I2/A03/Lien
16. Lien, H.-Y.: Effects of EFL individual learner variables on foreign language reading anxiety and metacognitive reading strategy use. Psychol. Rep. **119**(1), 124–135 (Aug. 2016). https://doi.org/10.1177/0033294116659711
17. Tien, C.-Y.: Factors of foreign language reading anxiety in a taiwan EFL higher education context. J. Appl. Linguist. Lang. Res. **4**, 48–58 (Jan. 2017)
18. Capan, A., Karaca, M.: A comparative study of listening anxiety and reading anxiety. Procedia—Soc. Behav. Sci. **70**, 1360–1373 (Jan. 2013). https://doi.org/10.1016/j.sbspro.2013.01.198
19. Saito, Y., Garza, T.J., Horwitz, E.K.: Foreign language reading anxiety. Mod. Lang. J. **83**(2), 202–218 (1999). https://doi.org/10.1111/0026-7902.00016
20. Zhao, A., Guo, Y., Dynia, J.: Foreign language reading anxiety: Chinese as foreign language in the United States. Morden Lang. J. **97**, 764–778 (Sep. 2013). https://doi.org/10.1111/j.1540-4781.2013.12032.x

21. Jalongo, M., Hirsh, R.: Understanding reading anxiety: new insights from neuroscience. Early Child. Educ. J. **37**, 431–435 (Apr. 2010). https://doi.org/10.1007/s10643-010-0381-5
22. Sabti, A.A., Mansor, Y.T.M.B.T., Altikriti, M.Q., Abdalhussein, H.F., Dhari, S.S.: Gender differences and foreign language reading anxiety of high school learners in an iraqi EFL context. Int. J. Appl. Linguist. Engl. Lit., **5**(5), Art. no. 5, (Sep. 2016), doi: https://doi.org/10.7575/aiac.ijalel.v.5n.5p.208
23. Bensalem, E.:Foreign language reading anxiety in the saudi tertiary EFL Context, (Oct. 2020)
24. Abubakar, M., Hairuddin, N.: An investigation of reading anxiety among efl young learners, (Sep. 2020)
25. Razak, N.A., Yassin, A.A., Moqbel, M.S.S.: Investigating foreign language reading anxiety among yemeni international students in malaysian universities. Int. J. Engl. Linguist. **9**(4), 83 (Jun. 2019). https://doi.org/10.5539/ijel.v9n4p83
26. Muhlis, A.: Foreign language reading anxiety among indonesian EFL senior high school students. Engl. FRANCA Acad. J. Engl. Lang. Educ., **1**(1), Art. no. 1, (Jun. 2017), doi: https://doi.org/10.29240/ef.v1i1.160
27. Wijayati, P.H., Mardianti, N., Murtadho, N.: The correlation between students' reading anxiety and their reading comprehension in esp context. Int. J. Lang. Educ. **5**(2), 15 (Jun. 2021). https://doi.org/10.26858/ijole.v5i2.15440
28. Al-Sohbani, Y.A.Y.: Foreign language reading anxiety among yemeni secondary school students. Transl. Stud. **06**(01), 9 (2018)
29. Faruq, A. Z. A.: Reading anxiety in english as a foreign language for undergraduate students in indonesia. TLEMC Teach. Learn. Engl. Multicult. Contexts, **3**(2), Art. no. 2, (Dec. 2019), doi: https://doi.org/10.37058/tlemc.v3i2.1275
30. Lu, Z., Liu, M.: An investigation of Chinese university EFL learner's foreign language reading anxiety, reading strategy use and reading comprehension performance. Stud. Second Lang. Learn. Teach. **5**(1), 65–85 (Jan. 2015). https://doi.org/10.14746/ssllt.2015.5.1.4
31. Jafarigohar, M., Behrooznia, S.: The effect of anxiety on reading comprehension among distance EFL learners. Int. Educ. Stud. **5**(2), p159 (Mar. 2012). https://doi.org/10.5539/ies.v5n2p159
32. Matsuda, S., Gobel, P.: Anxiety and predictors of performance in the foreign language classroom. System **32**, 21–36 (Mar. 2004). https://doi.org/10.1016/j.system.2003.08.002
33. Nazarinasab, F., Nemati, A., Mortahan, M.M.: The impact of foreign language reading anxiety and text feature awareness on university students' reading comprehension ESP Texts. Int. J. Lang. Linguist. (Sep. 2014). https://doi.org/10.11648/J.IJLL.S.2014020601.11
34. Hassaskhah, J., Joghataeian, S.: The role of foreign language reading anxiety in advanced learners foreign language reading comprehension, **3**, pp. 80–95, (Jan. 2016)
35. Ghonsooly, B., Loghmani, Z.: The relationship between EFL learners reading anxiety levels and their metacognitive reading strategy Use. Int. J. Linguist., **4**, (Aug. 2012), doi: https://doi.org/10.5296/ijl.v4i3.2021
36. Aghajani, M., Gholamrezapour, E.: Critical thinking skills, critical reading and foreign language reading anxiety in iran context. Int. J. Instr. **12**(3), 219–238 (Jul. 2019)
37. International School, Thai Nguyen University—Vietnam and L. Q. Dung, "Speaking Anxiety and Language Proficiency among EFL at A University in Vietnam," Int. J. Soc. Sci. Hum. Res., **03**(09), (Sep. 2020), doi: https://doi.org/10.47191/ijsshr/v3-i9-01
38. Trang, T. T. T., Moni, K., Baldauf, R. B.: Foreign language anxiety and its effects on students' determination to study English : to abandon or not to abandon?, p. 15 (2012)
39. Trang, T.T.T., Baldauf, R.B., Moni, K.: Investigating the development of foreign language anxiety: an autobiographical approach. J. Multiling. Multicult. Dev. **34**(7), 709–726 (Dec. 2013). https://doi.org/10.1080/01434632.2013.796959

40. Tran, T.T.T., Moni, K.: Management of foreign language anxiety: Insiders' awareness and experiences. Cogent Educ. **2**(1), 992593 (Dec. 2015). https://doi.org/10.1080/2331186X.2014.992593
41. Huberty, T.J.: Anxiety and depression in children and adolescents: Assessment, intervention, and prevention. p. 469 (2013) doi: https://doi.org/10.1007/978-1-4614-3110-7
42. J. Creswell, Guetterman, T.: Educational research: planning, conducting, and evaluating quantitative and qualitative research, 6th edition. (2018)

The Relationship between Nostalgia and Life Satisfaction in College: A Chained Mediation Model

Daosheng Xu[1,2] and Yiwen Chen[1,2(✉)]

[1] CAS Key Laboratory of Behavioral Science, Institute of Psychology, Beijing 100101, China
sheng8305@163.com, chenyw@psych.ac.cn
[2] Department of Psychology, University of Chinese Academy of Sciences, Beijing 100049, China

Abstract. Objective: In order to explore the multiple chain mediating effects of positive affect and self-esteem between nostalgia and life satisfaction in college students. Methods: 477 college students were investigated with Southampton Nostalgia Scale, Positive Affect Scale, Rosenberg Self-esteem Scale, Life Satisfaction Scale. Results: ① Nostalgia was significantly and positively correlated with positive emotion ($r = 0.12$, $p < 0.01$) and life satisfaction ($r = 0.13$, $p < 0.01$), also, significant positive correlations exists among other key variables. ② Structural equation models showed that, nostalgia could exert effects on life satisfaction ($\beta = 0.14$, $P < 0.05$), and also indirectly through the independent mediating effect of positive affect, the chain mediating effect of positive affect and self-esteem ($\beta = 0.10$, $P < 0.05$). Conclusion: Nostalgia could affect life satisfaction, not only through direct path, but also through the indirect path of multiple mediating effects of positive affect and self-esteem.

Keywords: Nostalgia · Life satisfaction · Positive affect · Self-esteem

1 Introduction

Analyzing domestic or international studies on nostalgia, there are studies that examine the effect of nostalgia on loneliness from the negative aspect [1], while some studies start from the positive aspect and focus on the effect of nostalgia on positive variables such as subjective well-being [2–4]. Also, there are studies that explore the effects of nostalgia on outcome variables from the perspective of mechanisms. For example, some studies have explored the mediating role of gratitude, and others have explored the effect of nostalgia on self-esteem or mood, which in turn affects subjective well-being. However, in the context of previous studies on nostalgia, it is clear that nostalgia does not simply produce experiences such as subjective well-being, but rather undergoes a more complex process, i.e., multiple mediations to achieve it.

In addition, regarding the conclusion between nostalgia and happiness, there are still some differences in previous studies, so does nostalgia bring about an increase or decrease in happiness, and if it is as the perspective of the former study about happiness enhancement, what is the whole enhancement route? Therefore, the specific path of enhancement is the focus of this research study. In this sense, this study will have important theoretical and practical significance to enable college students to develop a way to enhance their happiness through the widespread phenomenon of nostalgia.

With the above analysis, this study focuses on the effect of nostalgia on life satisfaction from a positive perspective and explores the mechanisms involved. In other words, to explore the role of self-esteem and positive emotions.

2 Theory Analysis and Research Hypothesis

2.1 The Direct Effect of Nostalgia on Satisfaction of Life

Studies in the literature on nostalgia have found that nostalgia is more prevalent in groups such as college students and that nostalgia occurs throughout their life span [5]. In one survey, 79% of college college student respondents reported that they experience nostalgia at least once a week [6]. Additional studies have found that nostalgia is prevalent among college students as well as factory workers and even adolescent children in China [5]. Another study showed that nostalgia is prevalent across the lifespan [7].

People look selectively at what happened in the past and will be specific to a certain time or a certain person, and in their memories, people feel sincere happiness and a sense of satisfaction. Nostalgia, as an experience with positive emotions, is of great value in maintaining and promoting the physical and mental health of individuals. A study by Zhou [5] found that, physiologically, nostalgia brings warmth to individuals and relieves pain in individuals from the physical level. At the psychological level, nostalgia is likened to a "reservoir" of positive emotions, which can act as a reservoir of positive emotions, i.e., bring about the experience of happiness. With above analysis of nostalgia, this study proposes the following hypotheses:

H1:Nostalgia positively predicts life satisfaction.

2.2 The Direct Effect of Nostalgia on Emotions

Nostalgia is an emotional experience, a complex emotional state that occurs when one misses the past. Early conceptualizations of nostalgia primarily explained it as a form of psychological disorder and as a negative state, such as a form of depression or homesickness. This may not be surprising, since many of the "symptoms" associated with "nostalgia" are negative (e.g., anxiety, sadness). However, there is also evidence that the concept of personal emotional warmth is more often associated with nostalgia because nostalgia is a predominantly positive experience and is of a bittersweet nature. In the field of empirical research, content analyses on nostalgia point out that nostalgia carries more positive affective elements [8], and many researchers view nostalgia as a positive, positive experience [9–11] and as an emotion associated with the self. A recent research program proposed nostalgia as a resource that enables people to acquire and maintain the feeling that their lives are meaningful [12]. Sederkides [11] and others thought that nostalgia have beneficial psychological functions. A series of subsequent studies have shown that nostalgia enhances positive emotions and can boost self-esteem and strengthen social bonds. In the study in question, nostalgia was manipulated by asking participants to recall a nostalgic or common autobiographical experience. Subsequently, subjects completed an assessment of the hypothesized function of nostalgia. Results found that nostalgia (compared to control) increased positive emotions, self-esteem, and enhanced social connectedness as assessed [8]. For example, by reducing loneliness and a stronger tendency to connect [13–15]. In conclusion, evidence supports the functional description of nostalgia, suggesting that it promotes psychological resources and well-being. Researchers have begun to consider nostalgia as a broader survival resource that helps people not only cope with death, but more broadly in their efforts to gain and maintain a sense of meaning in their lives. Accordingly, this paper proposes the following hypothesis:

H2:Nostalgia positively predicts positive emotions.

2.3 Chain Mediation Analysis of Nostalgia on Life Satisfaction

Nostalgia can elicit positive emotions, and there is a wealth of research in this area. The reason for this can be explained by the expanded construct theory of positive emotions: "When individuals have positive emotions, they are able to promote their thinking and expand the scope of their attention, cognition, and action". Based on this, these new ideas, experiences, and actions that the individual has can expand the individual's thinking and drive his or her actions. At the same time, this expansion also gives a certain opportunity, which is for the individual to use the resources in a sustainable way, and this resource can be stored for a long time and can be extracted at any time in later actions, as Fredrickson [16] has pointed out that this stored resource can improve the individual's chances of dealing with difficulties in later actions.

There is a relationship between resources and individual self-esteem, and this relationship can be elaborated in terms of the psychological mechanisms of self-esteem development. This contains the social comparison theory, and the self-enhancement theory both explain the relationship between an individual's use of resources and individual self-esteem from different perspectives. Social comparison theory states that "individuals build positive psychological resources through comparisons with others and through feedback from others, and this resource is an important source for generating self-esteem." Alternatively, self-enhancement theories, including self-protection theories, are articulated from the perspective of individuals seeking positive information about themselves to enhance and maintain self-esteem. Through feedback from others or acquired external information, individuals can store those positive resources through self-protection, and resources are able to keep self-esteem to a certain extent. Self-esteem to some extent can increase the level of life satisfaction. Wang Jiyu [17] surveyed 303 college students by means of a questionnaire to study the relationship between the two variables of self-esteem and life satisfaction, and the results of the study found that "students with high levels of self-esteem hold an affirmative and positive evaluation of real self-worth, show self-confidence in life, and have a positive upward emotional experience of self, which results in relatively high life satisfaction as well." Chen Lina and others [18] studied the relationship between life satisfaction and self-esteem using college students as subjects, and the results after statistical analysis found that "college students with high levels of self-esteem always have higher life satisfaction". Through the above findings, it is assumed that from the perspective of positive psychology, the positive qualities carried by nostalgia lead individuals to positive emotions, and positive emotions simultaneously expand to construct higher levels of self-esteem, and these increased levels of self-esteem are an important source of generating life satisfaction. Based on the above discussions and analysis, the following hypotheses are proposed:

H3:Self-esteem and positive emotions act as a chain mediator between nostalgia and life satisfaction.

Based on above analysis, this dissertation focuses on the relationship between nostalgia and life satisfaction and, at the same time, explores the chain mediating role played by positive emotions and self-esteem. The model diagram of this study is as follows (Fig. 1).

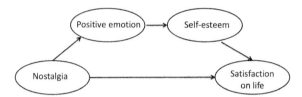

Fig. 1. Study structure

3 Subjective and Methods

3.1 Subjective

The subjects were selected from three universities in Shandong Province and Beijing, and 500 questionnaires were distributed to college students, and 477 valid questionnaires were obtained after the return check, with an effective rate of 95.4%. Among them, 231 were male students and 246 were female students. On the age variable, the maximum value was 25 years old, the minimum value was 18 years old, and the average age was 20.56 years old. On the grade variable, freshmen accounted for 35%, sophomores accounted for 41.8%, and juniors accounted for 23.2%. On the major variable, 46.5% of the students were in arts and 53.5% were in science. On the variable of family location, 56.5% were rural students and 43.5% were urban students. On the variable of whether they are poor students, 15.7% are poor students and 84.3% are non-poor students.

3.2 Tools

3.2.1 The Southampton Nostalgia Scale [19] (SNS) has a total of 5 entries to measure the frequency and tendency of individuals to be nostalgic and is scored on a scale of 1 to 7. In the present study, the internal consistency coefficient of the scale was 0.75.

3.2.2 Positive Affect Scale [20] (PAS) is a 9-item scale that measures the positive affect of individuals and is scored on a scale from 1 to 5. In this study, the internal consistency coefficient of positive affect was 0.892.

3.2.3 Rosenberg Self-esteem Scale [21] (SES) is a 10-item scale that measures the level of self-esteem of individuals. A scale of 1 to 4 was used. In the present study, the internal consistency coefficient of the scale was 0.764.

3.2.4 Satisfaction with Life Scale [22] (SWLS) is a 5-item scale that measures the individual's perceived evaluation of quality of life and is scored on a scale from 1 to 7. In the present study, the internal consistency coefficient of the scale was 0.773.

3.3 Data Processing

SPSS 17.0 was used for descriptive statistics analysis and correlation analysis, and AMOS 17.0 for multiple chain mediated effects analysis.

4 Outcomes

4.1 Validity and Common Method Bias Tests

Since this test was all filled out by college students themselves, in order to test the common method bias problem, this paper used Harman's one-way test for common method bias problem. After the test of factor analysis, it was found that the total explanation rate of the characteristic roots greater than 1 was 69.577%, among which the explanation rate of the first characteristic root was 24.322%, which was lower than the critical standard of 40%. This indicates that there is no serious common method bias problem in the data of this study.

The structural validity of the questionnaire was also explored in this study. The change in model fit after factor merging was tested, and the indicators used here include CFI and GFI below 0.05; and indicators such as NFI greater than 0.9 and RMSEA between 0.05 and 0.08. The specific method was to sum nostalgia, life satisfaction, self-esteem and positive emotion as a single factor, nostalgia, life satisfaction and self-esteem + positive emotion as three factors, nostalgia + life satisfaction, self-esteem + positive emotion as two factors, and nostalgia, life satisfaction, self-esteem, and positive emotion as four factors. After testing the indicators of the four-factor model were better than several other factor models, which proved that the variables in this study had good validity indicators. The final fit indices of the four-factor model were $\chi^2/df = 2.90$, CFI = 0.92, GFI = 0.91, NFI = 0.92, and RMSEA = 0.07. From the range of the above fit indices, the fit indices of the four-factor model were within the acceptable range, while the fit indices of the other factor models were not within the acceptable range. Therefore, from the above analysis, the potential common method bias does not have a significant effect on the relationship between the variables in this study.

4.2 Descriptive Statistical Analysis and Correlation Analysis

The results of the correlation analysis showed that the correlation between gender and self-esteem was significant and the correlation between gender and nostalgia was significant. There was a significant correlation between household registration and self-esteem variables. On the main variables, the independent variable nostalgia was significantly correlated with the dependent variable life satisfaction (r = 0.13, p < 0.01) and was significant at the 0.01; also, the independent variable, the dependent variable and the two main mediating variables showed significant correlations with each other and were significant at the 0.01 (Table 1).

Table 1 The correlation matrix between the variables (n = 477)

Variables	1	2	3	4	5	6
1. Gender	1					
2. Household registration	0.05	1				
3. Self-esteem	−0.15*	−0.09*	1			
4. Life satisfaction	−0.05	0.05	0.32**	1		
5. Positive emotions	−0.04	0.07	0.36**	0.44**	1	
6. Nostalgia	0.08*	−0.05	−0.01	0.13**	0.12**	1

Notes * P < 0.05, ** P < 0.01

4.3 Analysis of the Role of Multiple Chain Intermediaries

In this study, structural equation modeling analysis was used to verify the relationship between several variables. In the specific implementation, nostalgia was used as a predictor variable, while life satisfaction was used as an indicator variable, while positive emotions and self-esteem were used as mediating variables to construct a chain mediation model for testing. To present a better ratio between sample size and variables, this study used the item-structure balance method to pack the items of each scale in groups, in which nostalgia and life satisfaction were packaged into two item groups, and positive emotion and self-esteem were packaged into three item groups, respectively. Thus, the whole measurement model includes four latent variables and ten observed variables, and the model diagram is shown in Fig. 2.

From Fig. 2, from the fit indicators of the structural equation model, $\chi^2/df = 2.61$, which is less than 3, represents a good fit of the model. The other indicators: CFI = 0.98 and IFI = 0.98, the test criterion for both indicators is that if both indicators are greater than 0.90, it represents a good model fit, and in this study, both indicators are greater than 0.90, which further represents a good model fit. The third indicator, RMSEA, the test criterion for this indicator is that if the indicator is less than 0.05 represents a good model fit, and between 0.08 and 0.10, represents a fair fit. In this study, RMSEA = 0.05, which is in the range of good fit. Therefore, with the analysis of several indicators above, the model fit of this study is good.

Based on the good fit of the model, the next focus is on the path relationship between the variables. From the above figure, it can be found that the direct effect of nostalgia on life satisfaction is significant ($\beta = 0.15$, p < 0.05), while the indirect effect of nostalgia on life satisfaction is also significant ($\beta = 0.08$, p < 0.05), and among the indirect effects, nostalgia is mediated through positive emotions, as well as through the chain mediation of positive emotions and self-esteem.

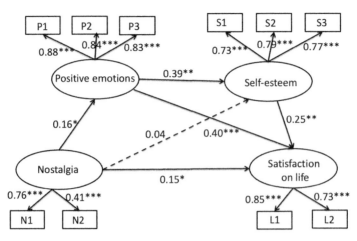

Fig. 2. Multiple mediation model between nostalgia and life satisfaction

5 Discussions

5.1 Discussion of the Direct Effect of Nostalgia on Life Satisfaction

Previous research on nostalgia has pointed out that nostalgia can store positive emotions and is a pleasant psychological state. As nostalgia research has become more rigorous, the idea that nostalgia is a self-relevant emotion and carries positive affective qualities has become more popular, with Sedikides [23] and others noting that there are self-relevant benefits to the positive quality of nostalgia. Belk [9] has empirically shown that "there is a positive relationship between the content of nostalgia and personal emotions", and Wildschut [6] has studied nostalgia as a "repository of positive emotions". They found that "subjects reported more positive emotions when they engaged in nostalgia". With above analysis, nostalgia's "storehouse of positive emotions" is an important source of life satisfaction for college students during their school years. In terms of its theoretical explanation, Hepper [24] pointed out that nostalgia, as a positive emotion pointing to the past, is a nostalgic experience in which people recall things that happened to them that are generally happy and meaningful, and as college students recall these positive and happy past events, these positive qualities can enhance positive emotions, which in turn can increase one's happiness and satisfaction.

5.2 Discussion on the Chain Mediating Effect of Positive Emotions and Self-esteem

Existing research perspectives addressing nostalgia suggest that an important role of nostalgia is its enhancing effect on self-esteem [6]. Research has also shown that nostalgia provides more direct benefits to the self by amplifying explicit self-esteem and buffers individuals' perceptions of life meaning, death-related anxiety, and related defenses from the negative effects of existential threat [25]. Extant research thus supports the validity of nostalgia constructs and highlights their self-related functions (e.g., supporting self-esteem). It has also been shown that a unique feature of nostalgia is the breadth of benefits it confers on the self. Nostalgia enhances social bonds and counteracts feelings of isolation [25], increases self-esteem [6], and buffers negative existential consequences reflecting personal mortality [13]. These benefits imply that nostalgia is a potential mechanism through which an individual can withstand various threats (e.g., social exclusion) to support the self. Thus, nostalgia may act similarly to other self-affirming resources in mitigating defensive responses to self-esteem threats [26, 27]. Past research has compared nostalgia and reflections to ordinary and past autobiographical events, such as in experiments showing that reflections to nostalgic events produce greater episodic self-esteem than reflections to ordinary past events [6].

The expanded-construct theory of positive emotions states that "positive emotions construct individual resources, which have value in themselves and play an important role in the physical and mental development of individuals. In this theory, individual resources include physical resources, interpersonal resources, and psychological resources. When positive emotions construct individual resources, they can enhance individuals' friendly behaviors, expand interpersonal resources, and enable subjective well-being to be enhanced." Considering previous empirical studies and the expanded construct theory of positive emotions, it can be found that starting from nostalgia can lead to positive emotions, and the positive qualities contained in positive emotions themselves are a direct source of life satisfaction. Therefore, in this sense, the mediation mechanism of nostalgia-positive emotion-life satisfaction in this study can be verified. In addition, according to the extended construct theory of positive emotion, after nostalgia triggers positive emotion, positive emotion expands individual psychological resources, such as self-esteem and self-efficacy, which are very important individual psychological resources, so in this sense, when positive emotion constructs individual psychological resources, such as self-esteem, it further enhances individual life satisfaction. Thus, the dual mediating effect of nostalgia-positive emotion-self-esteem-life satisfaction is also confirmed.

5.3 Research Significance

This study examines the effect of nostalgia on life satisfaction and concludes that both the direct and indirect paths of prediction are highly significant. This has important theoretical value for the inconsistency of previous findings on the prediction of nostalgia on subjective well-being. Throughout the previous studies, inconsistencies exist in the effects of nostalgia on subjective well-being, with some findings confirming the predictive effect of nostalgia, but others finding no predictive effect. The reasons for this contradictory result are mostly related to the understanding of subjective well-being. For example, Mingxia Wu [28] stated that subjective well-being is "the evaluator's holistic assessment of his or her quality of life according to his or her own criteria, and it is a comprehensive psychological indicator of an individual's quality of life." Life satisfaction is the degree of satisfaction judgments made by the evaluator about life, which belongs to the cognitive category of subjective well-being. In this sense, nostalgia may predict the cognitive component of subjective well-being (life satisfaction).

Also, the study has high practical significance. In this study, it was found that self-esteem plays an important role in the effect of nostalgia on life satisfaction, so the cultivation of college students' self-esteem and self-efficacy plays an important role in maintaining college students' life satisfaction. Self-esteem or self-efficacy is one's own evaluation of oneself. In the specific practice process, the cultivation of self-esteem of college students should start from themselves first, strengthen the friendly relationship with their classmates, especially cultivate good dormitory relationship, and the atmosphere of the whole group is an important way to cultivate self-esteem, in addition, in daily study, they should constantly set challenging goals for themselves, and through the continuous achievement of one goal In addition, you should constantly set challenging goals for yourself in your daily studies and improve your self-confidence by achieving them one by one. In addition, relying on the strength of the school, as the managers of the relevant departments of colleges and universities, they should actively provide ways for the cultivation of college students' self-esteem, first of all, they should recognize the role of counselors and class teachers in cultivating college students' mental health education, and the mental health level and character of counselors and class teachers have an important influence on college students' self-esteem. management team. In addition, relevant management departments of colleges and universities should provide corresponding activities for the healthy development of college students, such as various training courses or collective activities for various growth problems of college students, so that everyone can continuously discover and improve themselves in activities and courses.

6 Findings

6.1 Nostalgia has a significant positive correlation with life satisfaction, and the correlation between nostalgia and life satisfaction and the two main mediating variables is also significant.

6.2 Nostalgia has an indirect effect on life satisfaction by interacting with positive emotions and self-esteem, respectively. Meanwhile, nostalgia also has an indirect effect on life satisfaction through the effect of chain mediators between positive emotions and self-esteem.

6.3 The effect of nostalgia on life satisfaction has important theoretical significance and practical. For schools and people, consciously enhancing the level of life satisfaction through relevant management practices.

References

1. Xiulan, S.: Mechanism of the role of nostalgia and loneliness. Pract. Electron. **13**, 290 (2014)
2. Yue, L.: A study on nostalgia and its relationship with subjective well-being among college students. Southwest University, Chongqing (2012)
3. Rubo, C.: Mechanisms of Nostalgia's Influence on Subjective Well-Being—A Study of the Mediating Role of Gratitude. Jinan University, Giangzhou (2015)
4. Xiaoge, X.: An empirical study of the relationship between nostalgia and subjective well-being. Jinan University, Giangzhou (2013)
5. Zhou, X., Sedikides, C., Wildschut, T., Gao, D.-G.: Counteracting loneliness: on the restorative function of nostalgia. Psychol. Sci. **19**, 1023–1029 (2008)

6. Wildschut, T., Sedikides, C., Arndt, J., Routledge, C.: Nostalgia: content, triggers, functions. J. Pers. Soc. Psychol. **91**, 975–993 (2006)
7. Rosenberg, M.: Society and the Adolescent Self Image. Princeton University Press, Princeton, NJ (1965)
8. Wildschut, T., Sedikides, C., Routledge, C., Arndt, J., Cordaro, F.: Nostalgia as a repository of social connectedness: the role of attachment-related avoidance. J. Pers. Soc. Psychol. **98**, 573–586 (2010)
9. Batcho, K.I.: Nostalgia: a psychological perspective. Percept. Motor Skills, **80**, 131–143 (1995)
10. Davis, F.: Yearning for Yesterday: A Sociology of Nostalgia. Free Press, New York, NY (1979)
11. Sedikides, C., Wildschut, T., Baden, D.: Nostalgia: conceptual issues and existential functions. In: Greenberg, J., Koole, S., Pyszczynski, T. (eds.) Handbook of Experimental Existential Psychology, pp. 200–214. Guilford, New York (2004)
12. Juhl, J., Routledge, C., Arndt, J., Sedikides, C., Wildschut, T.: Fighting the future with the past: Nostalgia buffers existential threat. J. Res. Pers. **44**, 309–314 (2010)
13. Loveland, K.E., Smeesters, D., Mandel, N.: Still preoccupied with 1995: the need to belong and preference for nostalgia products. J. Consum. Res. **37**, 393–408 (2010)
14. Wildschut, T., Stephan, E., Sedikides, C., Routledge, C., Arndt, J.: Feeling happy and sad at the same time: Nostalgia informs models of affect. In: Paper Presented at the 9th Annual Meeting of the Society for Personality and Social Psychology, Albuquerque, New Mexico, USA (2008)
15. Zhou, X.Y., Wildschut, T., Sedikides, C., Chen, X.X., Vingei-hoets, AdJ. J.M.: Heartwarming memories: Nostalgia maintains physiological comfort. Emotion **12**,678–684 (2012)
16. Fredrickson, B.L., Coffey, K.A., Cohn, M.A., et al.: Open hearts build lives: positive emotions, induced through loving—kindness meditation, build consequential personal resources. J. Pers. Psychol. **95**(5), 1045–1062 (2008)
17. Jiyu, W.: The relationship between self-esteem and loneliness and life satisfaction among 303 college students. Chin. J. Sch. Doct. **24**(04), 254–256+259 (2010)
18. Lina, C.,Jianxin, Z.: General life satisfaction of college students and its relationship with self esteem. Chin Ment. Health J. (4), 46–48 (2004)
19. Routledge, C., Wildschut, T., Sedikides C, et al.: The power of the past: Nostalgia as a meaning-making resource. Meory **20**(5), 452–460 (2012)
20. Lin, Q., Xue, Z., Yanfei, W.: Revision of the positive affective negative affect scale (PANAS). Appl. Psychol. **14**(3), 249–254 (2008)
21. Feifei, C., Chongzeng, B., Mengfe, H.: Chinese Rosenberg positive self-esteem inventory reliability test. Adv. Psychol. **5**(9), 531–535 (2015)
22. Chengqing, X., Yuanli, X.: Reliability and validity of the Chinese version of the life satisfaction scale used in the population. China J. Health Psychol. **17**(8), 948–949 (2009)
23. Sedikides, C., Wildschut, T., Arndt, J., Routledge, C.: Self and affect: the case of nostalgia. In: Forgas, P. (ed.) Affect in social thinking and behavior: frontiers in social psychology, pp. 197–215. Psychology Press, New York (2008)
24. Hepper, E.G., Wildschut, T, Sedikides, C.: Down memory lane together: nostalgic interactions in close relationships. In: Paper Presented at the Poster Presented at International Association of Relationship Research Conference, Chicago, IL (2012)
25. Batcho, K.I., DaRin, M.L., Nave, A.M., Yaworsky, R.R.: Nostalgia and identity in song lyrics. Psychol. Aesthet. Creat. Arts **2**, 236–244 (2008)
26. Vess, M., Arndt, J., Routledge, C., Sedikides, C., Wildschut, T.: Nostalgia as a resource for the self. Self Identity **11**, 273–284 (2012)

27. Muehling, D.D., Sprott, D.E., Sultan, A.J.: Exploring the boundaries of nostalgic advertising effects: a consideration of childhood brand exposure and attachment on consumers' responses to nostalgia-themed advertisements. J. Adv. **43**, 73–84 (2014)
28. Mingxia, W.: Western theoretical developments on subjective well-being over the past 30 years. J. Dev. Psychol. **04**, 23–28 (2000)

Reform and Practice of Talent Training Model Based on Cold Chain Industry College

Huichuan Dai[1(✉)], Huihua Shang[2], and Yefu Tang[1]

[1] School of Management, Guangdong Province, Guangdong University of Science and Technology, Dongguan City 523070, China
86927520@qq.com

[2] School of Computer and Information Engineering, Guangdong Province, Hanshan Normal University, Chaozhou City 521041, China

Abstract. Industrial College has become an effective organizational form of cultivating applied talents. Exploring and constructing the talent training mode based on the industrial college is of great significance of deepening the cooperation between schools and enterprises, promoting the integration of industry, university and research, and improving the quality of talent training. Therefore, based on the analysis of the successful experience of applied talent training in home and abroad, this paper defines the construction idea of the talent training mode of the cold chain industry college. According to the practice and experience summary of the exploration of the new talent training mode by the cold chain industry college of Guangdong University of Science and Technology, we construct "3 + 1" talent training operation mechanism (which Let students study theoretical knowledge in school for the first three years, and enter the industrial class jointly established with enterprises in the fourth year to start practical learning), and analyze its theoretical basis, organization and leadership and curriculum system. It defines the key links such as learning centered teaching method, double qualified tutor team construction, practice base construction, assessment and evaluation reform, considers the construction of relevant systems, and summarizes the practical effect.

Keywords: Industry college · Talent training model · Learning-centered teaching method

1 First Section

Industrial College is an effective achievement of higher education organization innovation in recent years, an active exploration of school-enterprise cooperation in educating people, and a practical product of the deep integration of industry and education [1]. Many schools are actively exploring effective talent training modes of industrial colleges in combination with their own characteristics and the development needs and advantages of regional economy, and have achieved some successful experience. The school of management of Guangdong University of Science and Technology has established the cold chain industry college in cooperation with the provincial cold chain association in combination with the needs of regional economic development and the actual construction of logistics management specialty, and gradually explored and carried out the talent training in "3 + 1" industry innovation class. Therefore, it has become an urgent practical need that according to the actual needs of cold chain personnel training, building a scientific and rational mode of personnel training, deepening school enterprise cooperation, integration of production and teaching, and improving the quality of personnel training.

2 Ideas About the Construction of the Talent Training Model of the Cold Chain Industry College

Compare the more successful application-oriented talent training modes at home and abroad, such as German University of Applied Science and technology, British application-oriented talent training mode, Japanese application-oriented talent training mode, Shandong talent college talent training mode, animation industry college talent training mode, modern apprenticeship talent training mode [2–13]. It is not difficult to find their common ground. First, the goal of talent training is very clear, mainly to serve the local economy and cultivate high-quality applied talents that meet the needs of industrial development, have solid professional theoretical knowledge and high-level practical skills. Second, combine the professional characteristics of the colleges and universities and the actual needs of the local economy, deepen the integration of production and education, school-enterprise cooperation, strengthen the cultivation of students' practical ability, and have a high degree of corporate participation. Third, multiple parties jointly formulate talent training programs, innovative curriculum systems, and teaching methods. Fourth, reform and improve related management systems to ensure the smooth operation of new talent training models.

Therefore, the talent training model of the Cold Chain Industry College should use the Cold Chain Industry College as a platform for industry-education integration and school-enterprise cooperation to meet the needs of corporate professionals. We should establish a cooperative leadership agency for industrial colleges in secondary colleges of school. The school and cooperation enterprise should jointly formulate talent training programs, innovate curriculum systems, focus on ability improvement as the direction, and highlight practical curriculum construction. We should reform teaching methods, strengthen dual-qualified teachers team building, build a perfect practice training base by school-enterprise cooperation, and improve the quality of talent training in a sound operating mechanism.

3 "3 + 1" Talent Cultivation Operation Mechanism of Cold Chain Industry College

According to the actual situation of the operation of the Cold Chain Industrial College of Guangdong University of Science and Technology, the operating mechanism of the talent training mode as in Fig. 1. This mode aims to build a scientific and reasonable practical curriculum system and continuously improve the quality of talent training, based on the actual situation of the "3 + 1" industry innovation class, using the PDCA method to continuously improve the talent training program.

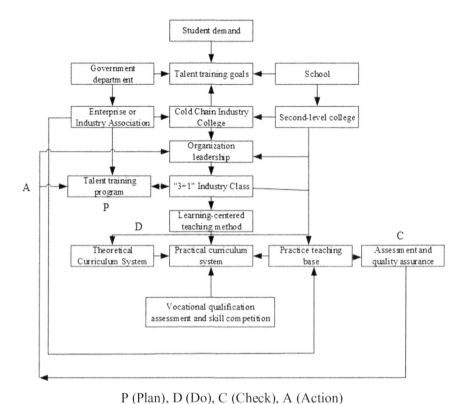

P (Plan), D (Do), C (Check), A (Action)

Fig. 1. Operation mechanism of talent training model of cold chain industry college.

3.1 Theoretical Basis of the New Operating Mechanism

The operating mechanism of the talent training mode of the Cold Chain Industry College refers to the composition of the elements in the process of talent training and the organic combination of each element. It is a dynamic system. Therefore, the theoretical basis of the new operating mechanism mainly includes system theory, total quality management theory, process method, etc. The rational use of the theoretical knowledge will help ensure the smooth implementation on the talent training mode of Cold Chain Industry College, and timely adapt to changes in market demand, respond promptly to problems arising from operation and take effective improvement.

3.2 Organizational Leadership

The core of the talent training model of the Cold Chain Industry College lies in the integration of industry and education, and school-enterprise cooperation. Among them, integration is the key. Generally speaking, it can be divided into three levels according to the level of integration. The first level is bonding. The second level is partial or partial integration. The third level is the deep integration of production, education and research. To ensure the smooth progress of integration and to enter the level of in-depth integration, a strong organizational leadership is indispensable. First, the leader should grasp the needs of all parties of a timely manner, the stakeholders have their own needs. The needs of all parties should be comprehensively considered and paid attention to and balanced to ensure the long-term development of the integration work. Second, the leader should be fully responsible and plan all tasks in the construction of new talent training model, determine talent training goals, formulate talent training plans, development plans and management systems. Third, the leader should evaluate, guide and supervise the specific implementation of new teaching model. Fourth, the leader should make other decisions on the construction of new teaching model.

3.3 Curriculum System

Curriculum system is the core of the reform of talent training mode. It runs through the whole teaching link, including theoretical curriculum system, practical teaching curriculum system and specific practical activities. "3 + 1" talent training model adopts the method of first concentrating theoretical learning, concentrating practice, summarizing and improving. In the specific implementation process, the curriculum system must be reasonably determined according to the professional standards and the basic law of ability formation. It should be determined in accordance with the principle of "ensuring foundation, paying attention to application, strengthening practice and highlighting characteristics", and should increase the proportion of practical courses to enhance ability training. Among them, theory courses can make full use of excellent online course resources such as China University MOOC and Love Course Network, and implement online and offline mixed education, so as to lay a solid foundation for the formation and improvement on practical ability and ensure the gradual improvement on students' ability level. Practice courses should be closely combined with theory courses, take the combination of remote video teaching and on-site practical teaching, and form a ladder according to the designed ability to build a practical curriculum system (Fig. 2).

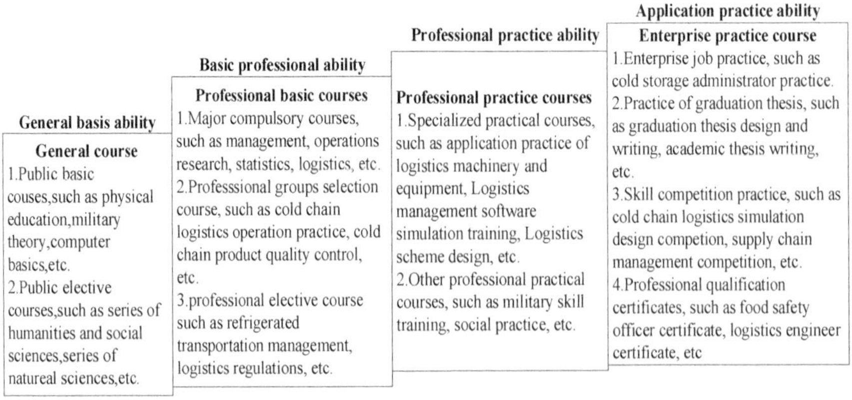

Fig. 2. Curriculum system based on the ability growth of cold chain talents.

Through the above courses, students are trained to love the motherland, care about society, protect the environment, love to labor, help students correctly master certain life skills, and investigate students' ability to comprehensively use the basic theories, basic knowledge and basic skills. It also focuses on gradually improving students' abilities, such as general basic ability, professional basic ability, professional practice ability when ensuring students to lay a solid professional theoretical foundation. It also adopts various forms such as professional skill competition, professional skill qualification certificate, internal and external practice base training, post practice of school enterprise cooperation units to comprehensively improve students' application practical ability.

A scientific and perfect curriculum system cannot be achieved overnight, which requires continuous summary and improvement. Therefore, it is necessary to gradually adjust and improve the whole curriculum system according to the PDCA cycle, so as to make it more adapt to the needs of development, and can continuously improve the training quality of applied talents.

3.4 Key Links of the New Operating Mechanism

Learning-centered teaching method. From the successful experience of applied talent training mode at home and abroad, effective teaching methods are an important factor to improve students' practical ability. Therefore, we should actively combine the reform of learning centered teaching methods carried out by the school, innovate teaching methods and teaching means, give full play to the creativity of students' self-learning, self-management and self-service, and mobilize the students' enthusiasm for active learning, active thinking and active innovation [14]. we should reform the traditional teaching mode centered on knowledge transfers, and realize the transformation from knowledge to ability training in a variety of effective methods such as project teaching method, case teaching method, situational teaching method, role-playing method, brainstorming method [15].

Team construction of dual-qualified tutors. "Dual teacher type" refers to excellent teachers from schools and technicians and engineers from the front line of enterprise production activities. The characteristic of Germany's "dual system" is to build a "dual classroom" between schools and enterprises through school-enterprise cooperation, cultivating professional basic skills, job practical skills and vocational skills.

The construction of the "double-qualified" teacher team comes from the adoption of an advanced and scientific teacher training model. It is reflected on three ways. One is to let teachers walk into the enterprise to understand the business management model and the employment environment of the enterprise. The teachers can learn practical skills in the enterprise in a targeted manner and apply them in the classroom. The other is to pass the national examination and obtain the corresponding qualifications. The third to let the teachers basically have all the requirements of a professional teacher by the accumulation of more practical experience, such as one year of guided teaching and one year of independent teaching [16].

Practice base construction. The practice base is an important foundation for the development of practice teaching. Schools, enterprises and government at all levels should attach great importance to the construction of practice bases, take effective measures, comprehensively utilize multiple resources, and improve the level of practice base construction [17]. At present, domestic enterprises and schools those use industrial colleges as a platform to carry out industry-education integration and school-enterprise cooperation all attach great importance to the construction of practice bases.

The logistics management major of Guangdong University of science and technology has always attached great importance to the construction of practice base. At present, the laboratory in the university is fully equipped. There are 8 practices and training rooms, including logistics training center, international logistics document laboratory, international logistics freight forwarding laboratory and international logistics business simulation training center. The total value of the equipment are 2.2645 million yuan, with an average student of 3400 yuan.

In order to better to meet the needs of practical training and teaching, this major has signed school-enterprise cooperation agreements with 15 enterprises including Dongguan Jiangnan agricultural products wholesale market management Co., Ltd, Carrefour Dongguan Commercial Co., Ltd. Hongwei Store, etc., to provide students with off-campus internships and condition. In-depth cooperation with Dongguan Jiangnan Market Management Co., Ltd., Dongguan Branch of SF Express, and other companies, relying on the company's cold chain logistics projects to provide practical teaching bases of logistics management major, jointly carry out scientific research projects, realize the complementarity of the resources of both parties, and collaborate to cultivate cold Chain logistics management talents.

Assessment and evaluation innovation. The implementation of "3 + 1" talent training mode poses new challenges to the assessment and evaluation of students' study. Because of the original planned courses cannot be studied and assessed as planned during the internship of the students in the enterprise, the traditional assessment and evaluation methods are obviously no longer applicable. For example, the part of the 2017 grade undergraduate teaching plan that is not taught in the seventh semester needs to adopt a variety of plans, such as credit exchange, online teaching, enterprise on-site teaching, which can be reasonably solved. In the future, we can also explore the ways and methods of practical ability assessment and evaluation by means of post skill certificate and credit mutual recognition, national vocational qualification certificate and credit mutual recognition. For example, Huawei Group and Guangzhou business school jointly build an industrial college, and incorporate Huawei international vocational certification system into the teaching process in the process of school enterprise cooperation, build the Pearson VUE international examination center on Huangpu district to enhance students' competitiveness through the symbiotic mode of "course certificate".

4 Construction of Relevant Regulations and Systems for Talent Training in Cold Chain Industry College

The training system is one of the important elements of the talent training model, it is also the basis of the normal operation of the new talent training model. Therefore, it is necessary to attach great importance of the construction of related systems. Judging from the successful experience of domestic colleges and universities, some corresponding systems will be specially formulated and established according to actual needs to ensure the normal development of related work of the industrial college, especially for some problems that arise from the daily operation process, should formulate targeted policies and measures to solve them.

4.1 "3 + 1" Talent Training Organization and Management System of Cold Chain Industry College

As the cold chain industry college involves the government, enterprises, schools and other parties. In order to carry out diversified cooperation smoothly, it is necessary to make clear agreements on the rights, responsibilities and interests of all parties, and fix them in the form of formal agreements and documents to form a system. There are usually three levels of rules and regulations to be established. One is the government level, such as the relevant policy documents formulated by the Ministry of education, Provincial Department of education, local government and other administrative institutions. The other is the relevant documents formulated by the school and enterprise level to promote the construction of industrial colleges, promote the implementation of industry education integration and school enterprise cooperation, and carry out effective management. For example, Guangdong University of Finance has formulated the measures for the construction and management of the Industrial College of Guangdong University of Finance, which defines the overall objectives, principles, construction contents and scope, establishment methods, organizational structure and management responsibilities, etc. Which will make the development of various work feasible. Third, it is necessary to formulate some specific implementation measures in the specific implementation process, such as the organization and management of students' internship in enterprises, internship content and practical skill certification, etc. These measures will help ensure the safety and smooth in the implementation process, ensure the realization of talent training objectives and the improvement on talent training quality.

4.2 "3 + 1" Talent Training Quality Assurance System of Cold Chain Industry College

The fundamental purpose of "3 + 1" talent training mode of cold chain industry college is to improve the quality of talent training and cultivate high-quality talents that meet the market demand and have strong innovation ability and practical ability. The improvement on talent training quality requires a perfect quality management system to ensure the efficient operation of talent training process and continuously improve the talent training process. Therefore, we must establish an effective teaching management system, teaching quality monitoring system and enterprise practice guidance norms according to the actual situation, so as to standardize all kinds of work, help to timely find and solve various problems of the new talent training mode, so as to ensure the quality of talent training and achieve the predetermined training objectives.

4.3 "3 + 1" Talent Cultivation Support and Incentive System of Cold Chain Industry College

As a new talent training model, schools, enterprises, teachers, students and other parties need to actively participate, especially the construction of practice bases and the improvement on teacher capacity, which are inseparable from the initiative of schools, enterprises, teachers and even students. As a result, it is necessary to establish an effective incentive support system to stimulate the enthusiasm and initiative of all parties. For example, as school teachers, they are easy to get used to traditional teaching methods and lack the motivation to actively improve their own practical ability. Schools can establish corresponding incentive policies to mobilize teachers' initiative, such as improving the treatment of dual-qualified teachers, Increasing opportunities for excellent evaluation and so on. Through effective incentives, all parties can take the initiative and make concerted efforts to jointly improve and perfect "3 + 1" talent training model, and achieve the goal of deep integration of production and education, high-level cooperation between schools and enterprises, and rapid improvement of talent quality.

5 Practical Effect

The cold chain industry college of Guangdong University of Science and technology that selects the best to establish a cold chain logistics innovation class through the two-way choice of enterprises and students, builds a cold chain logistics curriculum system to meet the needs of enterprises, and carry out teaching and enterprise practice. It carries out the learning of practical training courses based on the cold chain logistics laboratory to improve students' ability to combine basic theory and practice; and trains students' theoretical innovation ability through relevant professional competitions, such as logistics simulation design competition, logistics procurement competition, logistics scheme design competition, etc. Then, help students transform theoretical knowledge into practical work ability through the practice in cold chain logistics practice base. Further promote students' understanding and mastery of theoretical knowledge, improve students' comprehensive quality and increase development potential.

Since 2016, the number of students in the cold chain innovation class has been stable at about 30. After several years of improvement and promotion, the comprehensive ability and quality of students in the cold chain innovation class have improved significantly. The graduating students have also been recognized and reused by Dongguan Jiangnan agricultural products wholesale market management Co., Ltd, Shunfeng cold chain and other enterprises. After one or two years of post practice, many students have quickly stepped into enterprise backbone or middle-level leadership positions.

6 Conclusion

Generally speaking, the talent training mode of cold chain industry college aims to solve the problems of what kind of cold chain talents to cultivate, how to cultivate, and how to ensure the quality of talent training. Therefore, this paper analyzes the successful application-oriented talent training modes at home and abroad, summarizes the talent training experience of "3 + 1" industry innovation classes as the school of management of Guangdong University of science and technology, establishes the construction idea of talent training mode of cold chain industry college. The paper also builds "3 + 1" talent training operation mechanism of cold chain industry college, clarifies its theoretical basis, organization and leadership, curriculum system, and emphasizes the key links such as teaching methods, teacher construction, practice base construction. It builds the personnel training organization and management system, quality assurance system and support and incentive system are constructed to ensure the effective operation of the new mechanism. According to the actual operation effect, it is effective and feasible to continuously improve and make it continuously adapts to the needs of development.

Acknowledgments. 1. Grant sponsor: The "14th five year plan" of Guangdong Institute of higher education, 2021 higher education research topic, youth project. Grant No.: 21gqn06.

2. Grant sponsor:2021 university level teaching quality and teaching reform project of Guangdong University of Science and Technology. Grant NO.:GKZLGC2021055.

3. Grant sponsor:2021 university level teaching quality and teaching reform project of Guangdong University of Science and Technology. Grant NO.:GKZLGC2021022.

4. Grant sponsor:2019 university level "innovation and strengthening school project" (Research) of Guangdong University of Science and Technology. Grant NO.:GKY-2019CQYJ-14.

References

1. Yangfan, O.: Some thoughts on the integration of school-enterprise culture under the background of industrial colleges. J. Guangdong Vocat. Tech. Educ. Res. **4**, 55–56 (2019)
2. Fanglai, Z.: Comparative Study and Practice of the Training Model of Applied Talents Between China and Germany, 1st edn, pp. 6–10. Tsinghua University Press, Beijing (2014)
3. Tongwen, X., Yan, C.: Analysis and enlightenment on the training mechanism of applied talents in British Universities. J. Res. Higher Eng. Educ. **4**, 111–115 (2013)
4. Shuhan, L., Ling, B.: The theory and practice of school-enterprise cooperation applied talent training model, 1st edn, pp. 54–55. Nankai University Press, Tianjin (2014)
5. Cuilan, L., Yanke, Z.: Exploration and practice of applied talent training model in private undergraduate colleges, 1st edn, pp. 54–55. Shandong University Press, Jinan (2012)
6. Zeling, Z.: Study on the operation mechanism of the animation industry college under the background of the integration of production and education. J. Hunan Packaging **34**(185), 133–136 (2019)
7. Gengpu, Y., Chunli, W.: The reform and research of the talent training model for the integration of education, research and production in the "Animation and Animation" majors—taking Jilin Animation College as an example. J. Modern Educ. Sci. **3**, 142–145 (2013)
8. Huilan, G.: Exploration of the "Four Double Integration" talent training model in higher vocational colleges—taking the Industrial College of Zhongshan Vocational and Technical College as an example. J. Contemp. Vocat. Educ. **4**, 80–85 (2017)
9. Qin, Y.: Relying on industry colleges to comprehensively promote the reform of modern apprenticeship training model—taking Zhongshan Vocational and Technical College as an example. J. Vocat. Educ. Res. **10**, 18–24 (2019)
10. Liu, L., Zou, Y.: Exploration and practice on training mode of innovative network engineering talents. Int. J. Inform. Educ. Technol. **9**(6), 396–401 (2019)
11. Jixiang, Z., Yuezhou, Z.: Research on innovation and entrepreneurship talent training model for application-oriented university under perspective of collaborative innovation. Int. J. Inform. Educ. Technol. **9**(8), 575–579 (2019)
12. Jing'ai, L., Weiqing, L.: Research on talent training model of new applied undergraduate colleges. Int. J. Inform. Educ. Technol. **9**(9), 652–660 (2019)
13. Si, H., Xingliu, H., Tang, Y., Yang, Z.: Research on the construction of "Three Integration and Three Promotion" applied talents cultivate mode for automation major. Int. J. Inform. Educ. Technol. **9**(9), 666–670 (2019)

14. Yiping, J., Jingshan, W., Jin, C.: The exploration and successful practice of the innovation of the training model of compound talents—taking the intensive class of Zhu Kezhen College of Zhejiang University as an example. J. Res. Higher Educ. Eng. **3**, 132–136 (2012)
15. Zhong, S.H., Zhou, W.L.: Inquiry and experiential mixed teaching method is effective way to cultivate high-quality innovative talents. Int. J. Inform. Educ. Technol. **9**(9), 613–617 (2019)
16. Shuhan, L., Ling, B.: The Theory and Practice of School-Enterprise Cooperation Applied Talent Training Model, 1st edn, pp. 62–63. Nankai University Press, Tianjin (2014)
17. Lei, Y., Hua, W.C.: Strengthening scientific research ability of undergraduates, cultivating practical and innovative talents. Int. J. Inform. Educ. Technol. **3**(6), 648–650 (2013)

Study on the Growth Pattern of Middle-Level Vocational Skills

Xinqiang Meng(✉), Le Qi, and Mengyang Liu

Aviation Maintenance NCO Academy, Air Force Engineering University, Xinyang 464000, Henan, China
`qixiaole@buaa.edu.cn, 15373998415@163.com`

Abstract. Skilled professional personnel are the human resources that enterprises rely on for survival, and their professional competence level affects the long-term development of enterprises. The paper takes 86 vocational skilled talents as the sample, collects the data on the growth of professional ability such as education level, learning, and training, service experience, mentoring, organization and management, extracts the parameter indexes that can objectively reflect their professional ability quality, integrates the characteristics of various methods such as behavioral event interview method, questionnaire survey method, expert group discussion method and focus interview method for comprehensive analysis. Through combing and analyzing, the professional competency quality model of vocationally skilled personnel is constructed and its effectiveness for talent screening and assessment is verified. The model and its creation method can provide a useful reference for enterprises or related personnel.

Keywords: Career skills · Talent growth · Competency model

1 Introduction

After more than 40 years of reform and opening-up, China's economy has transformed from a low-value-added, labor-intensive development model to a technology-intensive one. Social and economic transformation needs a large number of craftsmen with professional skills to support, and vocational education is the cradle of skilled craftsmen. Middle-level vocational education is the starting point of training craftsman talents.

The competency model, also known as the "quality model", "competency model", "competency model", etc., refers to a combination of competencies that are required to be successful in a certain occupation. It is a combination of the competencies required to be successful in a particular career [1]. Specifically, it includes knowledge, skills, self-concept, traits, and motivation [2]. In 1973, American scholar David McClelland used the behavioral incident interview method to explore the competency characteristics of outstanding diplomats and to select outstanding diplomatic personnel. This event is generally regarded as the beginning of the development of the competency model. Based on this research, David McClelland argues that "the use of behavioral events as a basis for the selection of outstanding diplomats is a key element of the model." McClelland argued that "it is unreliable to predict future job success or failure using intelligence tests; there is no strong correlation between the results of intelligence tests and job success, and the relationship between the two must be determined on a case-by-case basis." Therefore, he advocates replacing intelligence tests with quality models as a new method for selecting suitable talents [3]. Domestic scholars' research on competency quality models began in the late 1990s, with specific research methods cited from developed countries combined with specific management practices of enterprises or projects [4].

The paper analyzed and researched the growth data of 86 professional skilled personnel from different departments and extracted the parameter indicators that can objectively reflect the professional competence quality of the sample from a large amount of information.

2 Analysis Process of the Growth Data of 86 Intermediate Vocational Talents

The first modeling approach is represented by David McClelland, which focuses on identifying the corresponding quality characteristics of individuals required for successful job performance, generally by selecting people with high performance in a certain area to conduct behavioral event interviews to refine their quality characteristics, and this approach is currently popular in Europe and the United States [5]. The second modeling method of the quality model is based on the organization's values. The premise of this modeling method is that the organization must have tested values, and through the organization's values construction model, the organization's values take root and are truly integrated into the quality requirements for employees. The third modeling method is based on the success factors of an industry. The key to this modeling method is to successfully identify the specific success factors of the industry and correspond to the quality requirements, and this method is now gradually applied to the construction of quality models [6]. The three modeling methods of the competency model are generally used interchangeably to complement each other. If only the third method is used, the reliability and validity of the data are not well guaranteed. The second method is generally used in mature enterprises or government agencies. The first and the third methods are combined to build the quality model.

Discovering quality characteristics through specific data collection behavioral patterns or performance has been recognized as one of the effective methods to collect quality elements. There are many different data collection methods for competency models, including behavioral event interviews, questionnaires, expert group discussions, and focus interviews, all of which have their unique characteristics [7].

The 86 intermediate vocational skills personnel selected for the paper were mainly the participants of vocational skills training, who had to take theoretical examinations, operational skills assessment, and comprehensive assessment before and after training, respectively, and submit technical summaries and corresponding supporting materials during their practice to the training center as required. Due to time reasons, it was not possible to conduct behavioral event interviews with each person. The research takes a compromise approach, using training lectures and appraisal opportunities, on the one hand, to follow up the entire learning and training process of the sample and communicate with the same book from all angles; on the other hand, relying on the appraisal station, the archived archival materials of the sample, including technical summaries and supporting evidence, were carefully reviewed and analyzed to extract data on the elements of professional competence [8]. At the same time, a questionnaire survey was conducted to extract the information elements of professional competence that are closely related to the growth and progress of vocationally skilled personnel.

2.1 Knowledge Structure - Educational Level

In the sample, there was 1 person with junior high school education, accounting for 1.2%; 1 person with high school education, accounting for 1.2%; 15 people with non-technical college education, accounting for 17.4%; 36 people with technical college education, accounting for 41.9%; 4 people with non-technical bachelor's degree, accounting for 4.7%; 7 people with technical bachelor's degree, accounting for 8.1%; and 22 people without graduation certificate, accounting for 25.6%. The distribution of sample education is shown in Fig. 1.

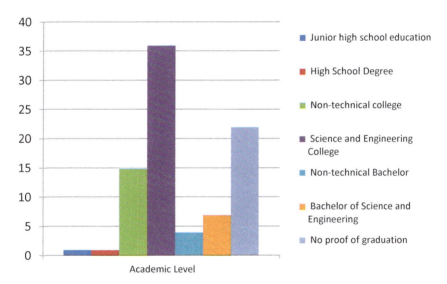

Fig. 1. Distribution of education in the sample.

As the self-study examinations become more and more strict, from the perspective of encouraging people to take self-study examinations to improve their qualifications, the way of obtaining qualifications is not distinguished from full-time or self-study examinations and is treated equally in the statistics [9]. According to the requirements of the industry, to obtain technician qualification, one must have a college degree or above in science and technology [1]. Half of the sample has not yet met the educational requirements, which will become a key factor limiting their ability to further develop in the industry. From the questionnaire survey, we can also see that the two words "education" appear frequently and have become a "distant concern" and "near worry" for many people.

2.2 Knowledge Structure - Learning and Training

In the sample, 8 people (9.3%) had one learning and training experience; 30 people (34.9%) had two learning and training experiences, and 48 people (55.8%) had more than three learning and training experiences. The learning and training situation of the sample is shown in Fig. 2.

Fig. 2. Sample learning training.

2.3 Duties and Responsibilities - Passing on the Help

The term "mentoring" refers to teaching recruits, commonly known as "apprenticeship." In the sample, 7 people (8.14%) had not taught recruits; 21 people (24.42%) had taught less than 3 people; 39 people (45.35%) had taught more than 3 people (excluding) and less than 10 people (excluding); 19 people (22.09%) had taught more than 10 people (including). The transference situation of the sample is shown in Fig. 3.

Leading and teaching new members and doing a good job of passing on the skills to others is an important means of passing on the business skills and industry spirit of the enterprise. The power of example is infinite, an excellent business backbone is often also an excellent master teacher, its talent training and construction and development of enterprises play an important leading role [2]. Those who have taught more than 3 people (not included) can be regarded as having richer teaching experience and playing a better role as a backbone for passing on the teaching.

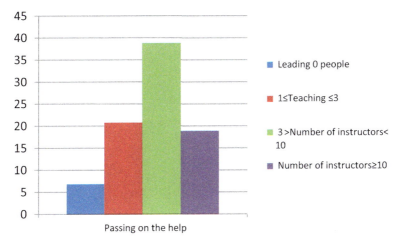

Fig. 3. Sample pass-along situation.

2.4 Competency - Organizational Management

The thesis considers the experience of having been a shift leader, team leader, or task leader as organizational management experience. In the sample, 24 people (27.9%) had organizational management experience; 62 people (72.1%) had no corresponding experience. It is easy to see that most vocational skilled personnel lack organizational management experience and do not have the corresponding organizational management ability, which is more consistent with reality. Many surveys found that personnel with organizational management experience are usually both business and technical backbone and have strong organizational management ability at the same time. This kind of personnel is the excellent representative and the best among the vocational skill talents that enterprises should focus on training and retaining. The organizational management experience of the sample is shown in Fig. 4.

2.5 Competency - Identify and Troubleshoot Problems

Of the sample, 78 (90.7%) had experience in identifying and troubleshooting problems; 8 (9.3%) had no corresponding experience. The discovery and troubleshooting problems of the sample are shown in Fig. 5.

The data showed that the majority of the sample had experience identifying and troubleshooting problems. However, the further investigation confirmed that most of these individuals had only identified and participated in troubleshooting problems. The specific analysis and troubleshooting of the problems were mainly organized and implemented by engineers. This is also consistent with the actual situation of the company, i.e., the analysis and troubleshooting of the problem rely mainly on the organization of engineers [10].

Fig. 4. Sample organization management experience.

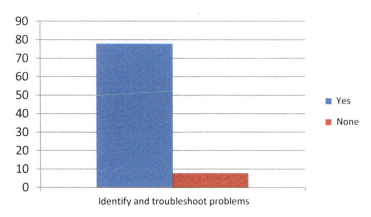

Fig. 5. Sample findings and troubleshooting problems.

2.6 Competency - Theoretical Research

The dissertation took writing research papers as an important reference for theoretical research ability. In the sample, 14 people (16.3%) had written research papers; 72 people (83.7%) had not written them. It can be seen that the vast majority of the sample had not written research papers and their theoretical research ability was weak. This is in line with the questionnaire survey that most of the vocational skilled personnel think they do not know the principle of equipment deeply and thoroughly, lack basic theoretical knowledge, and have weak research and analysis ability [4]. The theoretical research situation of the sample is shown in Fig. 6.

2.7 Competency - Job Competition

Enterprises usually recommend the selection of the backbone toppers with outstanding business ability to participate in various forms of job competitions [5]. Therefore, job

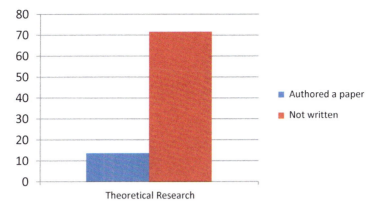

Fig. 6. Theoretical study of the sample.

competitions experience can be taken as an important reflection of the professional ability of vocational skill talents. In the sample, 26 people (30.2%) have participated in the job competition as a player; 60 people (69.8%) have not participated. Subjects' participation in the job competition is shown in Fig. 7.

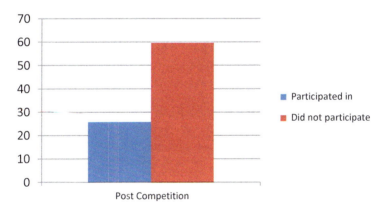

Fig. 7. Sample participation in job competition.

2.8 Honor - Access to Awards

In the sample, 50 people, or 58.1%, had received awards; 36 people, or 41.9%, had no corresponding awards. Some of the comrades who received awards were because they found larger fault problems, and some were because of outstanding work. In short, most of the comrades who received awards have made certain achievements in their work and have been affirmed by the enterprise, and their professional ability can be regarded as at least medium to high. The sample received awards as shown in Fig. 8.

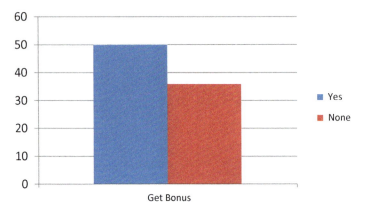

Fig. 8. The sample received awards.

3 The Establishment of Professional Competence Quality Model for Intermediate Vocational Skills Personnel

3.1 The Selection of Professional Competence Quality Indicators

After comprehensive data information from multiple channels and careful analysis and demonstration, the thesis divides the professional competence quality of intermediate vocational skills into four quality modules: knowledge structure, competence structure, the performance of duties and responsibilities, and honor [9]. Among them, the knowledge structure module includes three competency quality element indicators of education level, business regulations, and learning training; the competency structure module includes six competency quality element indicators of operational skills, theoretical research, organization and management, problem identification and troubleshooting, mentoring and job competition; the performance of duties and responsibilities module includes one competency quality element indicator of service experience; the honor module includes one competency quality element indicator of obtaining awards.

3.2 Modeling

With the above analysis, the thesis establishes a professional competence quality model for intermediate vocational skill personnel, as shown in Table 1.

4 Model Validation

Using the established professional competence quality model for intermediate vocational skills, personnel A and B were randomly selected from the sample for quantitative scoring, as shown in Tables 2 and 3. Person A scored 77 points and person B scored 25 points. The result of this scoring experiment is consistent with that of Delphi investigation method [11] shown in Table 4 symbolically. Therefore the proposed model can be used as an important reference for the selection and promotion of vocational skills personnel.

Table 1. Professional competency quality model for intermediate vocational skill personnel.

Quality module	Elemental indicators	Meaning	Quantification method
Knowledge structure	Academic level	Representing a certain learning ability and professional knowledge base, technician positions should have a college degree or above in science and technology	With a college degree or above in science and technology, 10 points, otherwise 0 points
	Business regulations	Theoretical knowledge of the business that the position should master as well as the corresponding regulations and systems	Relying on the appraisal station organization theory test obtained, 60 points and above to record business regulations qualified, 60 points below the record failed
	Learning training	An important way to broaden your horizons, optimize your knowledge structure, and enhance your professional ability and quality	10 points for more than 3 training experiences; 7 points for 2 training experiences; 5 points for 1 training experience; 0 points for none
Capability structure	Operating skills	The practical skills that should be available for the position	Relying on third-party organizations such as appraisal stations, 60 points and above to record operational skills qualified, 60 points or less to record unqualified
	Theoretical research	Writing research papers and conducting relevant theoretical research	10 points for writing more than 3 research papers; 7 points for 2 papers; 5 points for 1 paper; 0 points for 0 papers

(*continued*)

5 Discussion

Although the model proposed above is reasonablly valid in predicting the skill competence level and is much easier to use than traditionally used method such as expert scoring, the sampling population is small due to reasons that are being dealt with in

Table 1. (*continued*)

Quality module	Elemental indicators	Meaning	Quantification method
	Organizational management	Certain organizational coordination and management skills	10 points for organizational management experience, 0 points for none
	Identify and troubleshoot problems	Identify and troubleshoot problems can fully reflect the professional theoretical level and business ability of vocationally skilled personnel, representing a certain height of professional competence quality	10 points for being able to find and troubleshoot problems independently; 7 points for being able to find and participate in troubleshooting problems; 5 points for being able to find problems; 0 points for not finding problems
	Passing on the torch	Ability to teach and act as a mentor for new staff	10 points for teaching more than 10 persons; 7 points for teaching more than 3 persons and less than 10 persons; 5 points for teaching less than 3 persons; 0 points for teaching 0 persons
	Job competition	Participate in job competitions as a technical backbone	10 points for those who have participated in the competition and won; 5 points for those who have participated in the competition; 0 points for those who have not participated
Performing your duties and responsibilities	Experience	Important experience	10 points for more than 3 important postings, 7 points for 2 important postings, 5 points for 1 important posting, and 0 points for none

(*continued*)

succeeding researches. Cautions must be taken when critical assessment results are to be derived.

Table 1. (*continued*)

Quality module	Elemental indicators	Meaning	Quantification method
Honors	Get reward	Comrades who can receive awards usually have outstanding performance and at least good professional quality	Receiving annual awards or receiving the second prize of the post talent award above 20 points; receiving the third prize of the post talent award 15 points; receiving the team commendation, excellent employees, and other awards 10 points; no 0 points. The awards are not cumulative, and the points are recorded according to the highest award

Table 2. Comparison of professional competence of personnel A.

Quality module	Elemental indicators	A	
Knowledge structure	Academic level	Science and engineering college	10 points
	Business regulations	Qualified	
	Learning training	3 times	10 points
Capability structure	Operating skills	Qualified	
	Theoretical research	No papers written	0 points
	Organizational management	There are	10 points
	Identify and troubleshoot problems	Identify and participate in troubleshooting problems	7 points
	Passing on the torch	Leading 12 people	10 points
	Job competition	None	0 points
Performing your duties and responsibilities	Experience	More than 3 times	10 points
Honor	Get reward	2 times the excellent employee	20 points
Total		77 points	

Table 3. Comparison of professional competence of personnel B.

Quality module	Elemental indicators	B	
Knowledge structure	Academic level	Bachelor of science and engineering	10 points
	Business regulations	Qualified	
	Learning training	3 times	10 points
Capability structure	Operating skills	Qualified	
	Theoretical research	No papers written	0 points
	Organizational management	None	0 points
	Identify and troubleshoot problems	None	0 points
	Passing on the torch	No teaching	0 points
	Job competition	None	0 points
Performing your duties and responsibilities	Experience	1	5 points
Honor	Get reward	None	0 points
Total		25 points	

Table 4. Demonstration of Delphi Method

Expert NO.	Competence indicator 1	...		Competence indicator n
1	0.9		0.4	0.2
2	0.8		0.5	0.4
3	0.7		0.5	0.3
4	0.6		0.4	0.2
5	0.8		0.6	0.5
...	...		...	...
...	...		...	...
m				
Weighted average	Score 1	...		Score n

5.1 Conclusion

With a sample of 86 vocational skilled talents, various characteristic data of the sample were collected, from which the professional ability parameter indexes were extracted, and a variety of comprehensive analysis methods were applied to construct a professional ability quality model for vocational skilled talents. The experimental data analysis and comparison results showed that among many characteristic data, education, teaching experience, problem-solving experience, theoretical level, and job competition experience are the key independent variables affecting the quality of talent training. The results of this study will have some reference value for the cultivation of vocational skill talents in similar situations.

References

1. Gangtian, C.: Research and practice of promoting the professional cultivating ability in higher vocational colleges by ternary construction. Int. J. Inform. Educ. Technol. **9**(4), 314–317 (2019)
2. Wang, X.-M.: Effective implementation strategy of humanities quality education for higher vocational students in the age of new media era. Int. J. Inform. Educ. Technol. **9**(10), 762–766 (2019)
3. Chanchalor, S., Jitjumnong, K., Phooljan, P.: The effect of feedbacks on web-based learning modules for vocational students. Int. J. Inform. Educ. Technol. **9**(11), 767–771 (2019)
4. Xiao, B., Wei, M., Mingshe, D.: Construction and empirical analysis of the evaluation index system for majors set at transportation vocational colleges. Int. J. Inform. Educ. Technol. **9**(11), 778–783 (2019)
5. Wahyuni, D.S., Agustini, K., Ariadi, G.: An AHP-based evaluation method for vocational teacher's competency standard. Int. J. Inform. Educ. Technol. **12**(2), 157–164 (2022)
6. Gutiérrez, I., Sánchez, M.M., Castañeda, L., Prendes, P.: Learning e-learning skills for vocational training using e-learning: the experience piloting the (e)VET2EDU project course. Int. J. Inform. Educ. Technol. **7**(4), 301–308 (2017)
7. Mohamad, M.M., Sulaiman, N.L., Salleh, K.M., Sern, L.S.: Innovative invention skills and individual competency model for vocational education. Int. J. Inform. Educ. Technol. **7**(7), 514–517 (2017)
8. Ayub, H.: Parental influence and attitude of students towards technical education and vocational training. Int. J. Inform. Educ. Technol. **7**(7), 534–538 (2017)
9. Lai, F.-P., Liao, C.-W., Shih, C.-L., Chin-Chang, W.: A study on learning motives and learning effects of vocational high school students to energy technology education integrated into project course. Int. J. Inform. Educ. Technol. **5**(6), 418–424 (2015)
10. Chiang, C.L., Lee, H.: The effect of project-based learning on learning motivation and problem-solving ability of vocational high school students. Int. J. Inform. Educ. Technol. **6**(9), 709–712 (2016)
11. Roma construction of horizontal framework of "evidential reasoning" ability—based on Delphi survey. Chem. Teach. **3**, 6 (2021)

Author Index

A
Arones, Maritza, 135

B
Babayants, Carina, 34

C
Cai, Qihang, 88
Chandran, Sarath, 47
Chauca, Carmen, 135
Chen, Yiwen, 182
Cheng, Eric C. K., 3
Chernysheva, Anastasiia, 134
Curro-Urbano, Olga, 135

D
Dai, Huichuan, 193
Dang, Van T. T., 168

F
Fritsche, Katrin, 111
Fu, Xiuli, 102, 157

K
Khlopotov, Maksim, 34
Koppel, Maurice ten, 69

L
Le, Qi, 203
Liang, Ronghua, 102, 157

Libbrecht, Paul, 69

M
Mengyang, Liu, 203
Miao, Xin, 47
Monroe, Samantha, 47

N
Nadaf, Ali, 47
Nguyen, Trung, 168
Niu, Lei, 88

P
Peirong, Q. I., 20
Peng, Kunling, 125
Phun-Pat, Ynés, 135

Q
Qi, Yuansheng, 102, 157
Qian, WANG, 20

S
Schlippe, Tim, 69, 111
Shang, Huihua, 193
Shnaider, Polina, 34
Shuzhen, L. U. O., 20
Stierstorfer, Quintus, 69
Sun, Ying, 111

T
Tang, Yefu, 193

W

Wang, Mengmei, 147
Wang, Tianchong, 3
Wang, Yaohan, 125
Wölfel, Matthias, 111
Wu, Qi, 125

X

Xinqiang, Meng, 203
Xu, Daosheng, 182

Z

Zhang, Yongbin, 102, 157
Zheng, Yanying, 102, 157